CAMBRIDGE LIBRARY COLLECTION

Books of enduring scholarly value

Life Sciences

Until the nineteenth century, the various subjects now known as the life sciences were regarded either as arcane studies which had little impact on ordinary daily life, or as a genteel hobby for the leisured classes. The increasing academic rigour and systematisation brought to the study of botany, zoology and other disciplines, and their adoption in university curricula, are reflected in the books reissued in this series.

Makers of British Botany

First published in 1913, this volume reproduces a series of lectures on influential botanists delivered in 1911 in the Botanical Department of University College, London. The subjects of these biographies include Sir William Hooker (1785–1865), the first Director of Kew, and John Ray (1627–1705), considered the founder of scientific botany in Britain. The biographies are written by distinguished botanists of the period, over half of them holding a university professorship and membership of the Royal Society. Edited by F.W. Oliver (1864–1951), Professor of Botany at University College, London, from 1890 to 1929, these essays provide information on the lives, and discuss the scientific contributions, of each botanist, each contributor specialising in the same area of botany as the subject. The wide range of subjects covered demonstrates the development of key botanical concepts and the growth of scientific botany in the eighteenth and nineteenth centuries.

Cambridge University Press has long been a pioneer in the reissuing of out-of-print titles from its own backlist, producing digital reprints of books that are still sought after by scholars and students but could not be reprinted economically using traditional technology. The Cambridge Library Collection extends this activity to a wider range of books which are still of importance to researchers and professionals, either for the source material they contain, or as landmarks in the history of their academic discipline.

Drawing from the world-renowned collections in the Cambridge University Library, and guided by the advice of experts in each subject area, Cambridge University Press is using state-of-the-art scanning machines in its own Printing House to capture the content of each book selected for inclusion. The files are processed to give a consistently clear, crisp image, and the books finished to the high quality standard for which the Press is recognised around the world. The latest print-on-demand technology ensures that the books will remain available indefinitely, and that orders for single or multiple copies can quickly be supplied.

The Cambridge Library Collection will bring back to life books of enduring scholarly value (including out-of-copyright works originally issued by other publishers) across a wide range of disciplines in the humanities and social sciences and in science and technology.

Makers of
British Botany

A Collection of Biographies by Living Botanists

EDITED BY F.W. OLIVER

CAMBRIDGE
UNIVERSITY PRESS

CAMBRIDGE UNIVERSITY PRESS

Cambridge, New York, Melbourne, Madrid, Cape Town, Singapore,
São Paolo, Delhi, Dubai, Tokyo, Mexico City

Published in the United States of America by Cambridge University Press, New York

www.cambridge.org
Information on this title: www.cambridge.org/9781108016025

© in this compilation Cambridge University Press 2010

This edition first published 1913
This digitally printed version 2010

ISBN 978-1-108-01602-5 Paperback

MAKERS OF
BRITISH BOTANY

CAMBRIDGE UNIVERSITY PRESS
London: FETTER LANE, E.C.
C. F. CLAY, Manager

Edinburgh: 100, PRINCES STREET
London: WILLIAM WESLEY & SON, 28, ESSEX STREET, STRAND
Berlin: A. ASHER AND CO.
Leipzig: F. A. BROCKHAUS
New York: G. P. PUTNAM'S SONS
Bombay and Calcutta: MACMILLAN AND CO., Ltd.

John Hutton Balfour (1878)

MAKERS OF
BRITISH BOTANY

A COLLECTION OF BIOGRAPHIES
BY LIVING BOTANISTS

Edited by

F. W. OLIVER

Cambridge :
at the University Press
1913

Cambridge:

PRINTED BY JOHN CLAY, M.A.

AT THE UNIVERSITY PRESS

PREFACE

THE origin and scope of the present book will be found fully indicated in the *Introduction*, so that it is not needful to refer to them here. One change in the original scheme of the work has been made during its passage through the press, viz. the inclusion of an additional chapter from the pen of Prof. F. O. Bower dealing with the life of the late Sir Joseph Hooker. Our veteran botanist passed away on Dec. 10, 1911, in his 95th year, and in him botany loses its outstanding personality as well as its principal link with the past. The history of botany in this country during the Victorian period, when it comes to be written, must of necessity be woven around the life of this great man.

For the excellent index to the book the reader is indebted to Dr E. de Fraine, whose care and good judgment in this matter will be fully appreciated.

F. W. O.

October, 1912

CONTENTS

LIST OF ILLUSTRATIONS

INTRODUCTION

THE present volume represents in somewhat expanded form a course of lectures arranged by the Board of Studies in Botany of the University of London and delivered during the early part of 1911 in the Botanical Department of University College, London.

These lectures, which were ten in number, were widely attended by advanced and post-graduate students of the University and others interested in the subject.

The ten lectures comprised in the course were delivered by various botanists, the lecturer in each case being either a worker in the same field as, or in some other way having a special qualification to deal with, his allotted subject.

In view of the interest aroused by their delivery the hope found wide expression that the lectures might be issued in book form. At the time when the arrangements were being made for publication the University of London Press had not yet reached the publishing stage, so hospitality had to be sought elsewhere. That the book is issued from the Cambridge University Press is largely due to the good offices of Prof. A. C. Seward.

In consenting to publish *The Makers of British Botany* the Cambridge University Press suggested that some additional chapters should be prepared so that the work might be more fully representative. This has been done so far as was possible in the time available.

The sixteen chapters forming the book include (1) the ten lectures, which are printed essentially as they were delivered, (2) six additional chapters specially written under the circumstances just mentioned. As a rule each chapter will be found to deal with a single Botanist; with the exception of the first and

last chapters. In the former Prof. Vines has linked together
Morison and Ray, the founders of Systematic Botany in this
country, whilst in the last Prof. Bayley Balfour has expanded
what was originally intended as a sketch of his father, the late
Prof. J. Hutton Balfour, into a very interesting account of his
precedessors in the Edinburgh chair from the year 1670 almost
down to the present time.

The subjects treated, the authors and the order of arrangement are as follows :—

Subject	Born	Died	Author
*Robert Morison	1620	1683	}Prof. S. H. Vines, F.R.S.
*John Ray	1627	1705	
*Nehemiah Grew	1641	1712	Mrs Arber
*Stephen Hales	1677	1761	Francis Darwin, F.R.S.
John Hill	1716	1775	T. G. Hill
*Robert Brown	1773	1858	Prof. J. B. Farmer, F.R.S.
*Sir William Hooker	1785	1865	Prof. F. O. Bower, F.R.S.
*The Rev. J. S. Henslow	1796	1861	The Rev. Prof. Geo. Henslow
John Lindley	1799	1865	Prof. Frederick Keeble
*William Griffith	1810	1845	Prof. W. H. Lang, F.R.S.
*Arthur Henfrey	1819	1859	Prof. F. W. Oliver, F.R.S.
*William Henry Harvey	1811	1866	W. Lloyd Praeger
The Rev. Miles Berkeley	1803	1889	George Massee
Sir Joseph Gilbert	1817	1901	Prof. W. B. Bottomley
*William Crawford Williamson	1816	1895	Dr D. H. Scott, F.R.S.
Harry Marshall Ward	1854	1905	Sir William Thiselton-Dyer, K.C.M.G., F.R.S.
The Edinburgh Professors	1670	1887	Prof. I. Bayley Balfour, F.R.S.

* Was the subject of a lecture in the University Course.

The first three chapters deal with the founders of British
Botany, MORISON and RAY in the systematic field, GREW,
the plant anatomist, and HALES the physiologist. These are
pioneers and the names of Ray, Grew, and Hales must always
remain illustrious in the annals of Botanical Science.

JOHN HILL, with all his versatility, belongs to another plane,
but his inclusion here is justified on historical grounds, by the
prominent part he played in making known the method of the
great Swedish systematist Linnaeus, a method which took deep
root and gave an immense stimulus to systematic studies in this
country.

In ROBERT BROWN we have the greatest botanist of his day, for thirty years keeper of the Botanical Department of the British Museum. It is doubtful if any greater intellect than Brown's has ever been devoted to the service of Botanical Science.

SIR WILLIAM HOOKER was the first Director of Kew, and under his genial administration the foundations of that great institution were most truly laid. Born under the star of Linnaeus, his own researches lay in the systematic field—more especially among the Ferns and Bryophytes.

J. S. HENSLOW was for many years Professor of Botany at Cambridge, but it is his life as Rector of Hitcham in Suffolk that finds special prominence in the interesting Memoir which formed the subject-matter of his son's lecture. The account given of his educational methods will be read with interest in these days when " Nature Study " has been sprung on the world as a new thing.

JOHN LINDLEY was a man of the most amazing energy and his scientific output was prodigious. Though he attained high distinction in many fields of Botany, being an accomplished Systematist and Palaeobotanist, probably his greatest service was on the scientific side of Horticulture. Considering the scale of production, the work of Lindley maintains a remarkably high level. It is recorded of him that he never took a holiday till he reached the age of 52. His was the dominant personality in Botany of the early and mid-Victorian era.

WILLIAM GRIFFITH had the energy and power of endurance of Lindley, under whose influence he came. Trained to the practice of medicine he took service under the East India Company where he was able to devote the priceless intervals between his official duties to botanical travel, collecting, and the morphological investigation of Indian plants. The results of his brief but remarkable career are embodied mainly in his voluminous illustrated notes which were published posthumously in 1852. The name of Griffith has been happily linked with that of Treub, his brilliant successor in our own times.

ARTHUR HENFREY belonged to a very different type. Compelled by ill-health to the life of a recluse, his short life was

mainly devoted to making known in England the great discoveries of the Hofmeisterian epoch. To Henfrey belongs the
credit of being the first of our countrymen to recognise the full
significance of the new morphology, the general recognition of
which, however, he did not live to see. Henfrey was an extremely competent all-round Botanist whose single-minded
devotion to his subject should not be allowed to fall into
oblivion.

WILLIAM HENRY HARVEY is a representative of a numerous
class among the followers of Botany in this country. A man
of great personal charm and high culture, he was attracted into
the subject from the love of collecting. His special field was
that of the Marine Algae, in which he stood unrivalled. Harvey
was an exquisite delineator of the seaweeds of which he was so
enthusiastic a student. The memoir, based on his journals and
letters, which was published shortly after his death, is a book
well worth reading for its intimate sketches of the naturalists of
his day and the vivid notes on his extended travels in the
colonies and elsewhere.

MILES JOSEPH BERKELEY, like his contemporary Harvey,
was a cryptogamic botanist. He was a voluminous contributor
to the systematic literature of the Fungi over a period of fifty
years, as well as being a pioneer in the field of plant pathology.
The systematic collections accumulated during his long life form
one of the glories of the Kew Herbarium.

SIR JOSEPH HENRY GILBERT'S outlook on plants was
entirely different from that of any of the foregoing. He regarded the plant essentially as the chemical offspring of the
environment to which it was exposed. His life was devoted
to the study of soils and crops in conjunction with Sir John
Lawes. To these classic investigations carried out at Rothamsted, Gilbert brought the trained skill of the chemist.

WILLIAM CRAWFORD WILLIAMSON was a great all-round
naturalist of the Victorian period whose work as a Zoologist
gained him high distinction long before his attention became
seriously concentrated upon his famous studies into the structure
of the fossil plants of the Coal Measures. Though these researches were pursued without any marked contemporary

encouragement, at any rate until the closing years of his life, the field in which Williamson was so enthusiastic a pioneer has since his time been generally recognised as of the first importance—more especially in its bearing upon the pedigree of the vegetable kingdom. To-day, no branch of Botany has more recruits or is more vigorously pursued in this country than that of Palaeobotany, and so long as the science remains will the memory of Williamson be green.

HARRY MARSHALL WARD belongs to a generation younger than any of the foregoing. His student days coincided with the renaissance of Botany in England in the seventies of the last century, and coming under the influence of Huxley, Thiselton-Dyer, Vines and others, Ward early revealed himself as an ardent investigator. For twenty-five years he devoted his remarkable energies to a series of connected researches bearing broadly on the nutrition of the Fungi and allied organisms with especial reference to the relationships between host and parasite. The notice of his career which appears in this volume is from the pen of Sir William Thiselton-Dyer. Recently printed in the Obituary Notices of Fellows of the Royal Society, we are indebted to the courtesy of the author and of the Council of the Royal Society for permission to include it here.

In a book like the present, the work of a large number of distinct contributors, it is evident that no continuous or homogeneous treatment of the history and progress of Botany in this country is possible. Judged even as a series of essays or studies of representative men, *The Makers of British Botany* will not escape criticism, so long as special reference to the work of Priestley, Cavendish and Sénébier finds no place in its pages, not to mention such obvious omissions as Knight, Daubeny and Bentham. These omissions have not been deliberate and it will no doubt be possible to repair them should a second edition of the work be called for. The case of Charles Darwin is different. Apart from the work for which he is most famous, Darwin was a great investigator of the movements of plants and of the biology of flowers. As this aspect of Darwin's work has received adequate treatment in the recent centenary volume published by

the Cambridge University Press[1], it has not seemed necessary on
the present occasion to traverse the ground again.

   The reader of *The Makers of British Botany* will judge, and
we think rightly, that Botany has had its ups and downs in this
country. At the end of the seventeenth century England was
contributing her full share to the foundation and advancement
of the subject. In the field of Systematic Botany Ray, at any
rate, left his permanent influence as a taxonomist, whilst in
Plant Anatomy, the offspring of the newly invented microscope,
Grew divided the honours with his brilliant contemporary
Malpighi. A few years later Stephen Hales was carrying out
the famous experiments which are embodied in his *Vegetable
Staticks*, entitling him to be justly regarded as the Father of
plant physiology. Notwithstanding so admirable a beginning,
the next century was almost a blank. The essay on John Hill
serves to illustrate the sterility of this period. The dominant
influence in Botany in the eighteenth century was that of
Linnaeus, whose genius as a taxonomist gave the most wonderful
impulse to the study of Botany that it has ever received. Shorn
of its accumulated dead-weight of nomenclature, the simplified
Botany of Linnaeus took deep root in this country and here for
a century it reigned supreme as a source of inspiration. Fed on
unlimited collections of plants from all parts of a growing
Empire, it is hardly surprising that a great British school of
Systematic Botany led by Robert Brown, the Hookers, Lindley
and Bentham should have arisen. What is remarkable is the
almost exclusive persistence of this branch of Botany for more
than a generation after the establishment and recognition of
other departments on the continent of Europe. Whilst we
made a shrine for the Linnaean collections, so far as we were
concerned Grew and Hales might never have lived; even the
rational and scientific morphology created by Hofmeister in
the forties of last century failed to deflect us from our course!
   It was only in the later seventies that the New Botany came
to England, whither it was imported from Germany. For a
while, as was to be expected, our Universities were kept busy

[1] *Darwin and Modern Science.*

in training students in the modern work and in the conduct of investigations in the fields thus opened. With acclimatisation certain distinctive branches which may be regarded as characteristic have come to the front. These include more especially the study of anatomy in its phylogenetic aspects with which is closely linked that of the palaeozoic fossils, so richly represented in some of our coal-fields as to constitute a virtual monopoly. The present wide-spread revival of interest in palaeobotany is in no small measure attributable to Williamson, who, in spite of discouragement, kept the subject alive till the modern movement was firmly enough established to take up his work. Another productive field has been that of the nuclear cytology of both higher and lower plants, whilst physiology, especially on the chemical side, has attained pre-eminence. On present indications it is to be expected that in the near future physiology will receive much more attention than hitherto, partly as an inevitable reaction from the field of pure structure, and partly because of its fundamental importance in relation to agriculture. Nor is this the only branch that should be greatly stimulated by the forward movement in Agriculture that is now just beginning to be felt. The science of plant breeding, too long neglected by the countrymen of Darwin, has been pursued with much success for a decade, and has already reached the " producing stage " in respect of new and improved races of agricultural plants.

The youngest branch of Botany is Ecology or the study of vegetation in relation to habitat—particularly soil in its widest sense. This department deals with the recognition and distribution of the different types of plant community in relation to topography and the factors—chemical, physical and biologic— which determine this distribution. Ecology has the great merit of taking its followers into the field, where they are confronted with a wide range of problems not hitherto regarded as strictly within the province of the botanist. At the same time it exacts the most critical acquaintance with the minutiae of the taxonomist, so that a new sphere of usefulness is opened to the systematist. Ecology should have a great part to play in helping to break down the frontiers which have too long tended to separate Botany from the other sciences, and the maintenance of which is not in the true interests of the subject.

ROBERT MORISON AND JOHN RAY
1620—1683   1627—1705
By SYDNEY HOWARD VINES

Early systems of classification—Theophrastus—the Herbalists—Cesalpino's
De Plantis—Caspar Bauhin's *Pinax Theatri Botanici*—MORISON—
narrative—Botany at Oxford—the garden established—Jacob Bobart
the elder—Morison's *Historia Plantarum*—completion by the younger
Bobart—personal characteristics—Morison's works—the *Praeludia*—the
Hallucinationes—the *Dialogus*—principles of method in his *Plantarum
Umbelliferarum Distributio Nova*—posthumous publication of System—
indebtedness to Cesalpino—Linnaeus' estimate of Morison—RAY—narrative—first attempt at a System—quarrel with Morison—the *Methodus
Nova*—Dicotyledones and Monocotyledones—Linnaeus' criticisms—
later Systems—the French school—Morison and Ray compared.

THE literature of Botany can be traced back to a quite
respectable antiquity, to the period of Aristotle (B.C. 384—322)
who seems to have been the first to write of plants from the
truly botanical point of view. Unfortunately, his special treatise
on plants—θεωρία περὶ φυτῶν—is lost ; and although there are
many botanical passages scattered throughout his other writings
(which have been collected by Wimmer, *Phytologiae Aristotelicae
Fragmenta*, 1836), yet none of them gives any indication of what
his ideas of classification may have been. An echo of them
is perhaps to be found in the works of his favourite pupil,
Theophrastus Eresius (B.C. 371—286), who among all his fellows
was the most successful in pursuing the botanical studies that
they had begun under the guidance of the master. Theophrastus
left behind him two important, though incomplete, treatises on
plants, the oldest that have survived : the more familiar Latin
titles of which are *De Historia Plantarum* and *De Causis*

Plate I

ROBERTUS MORISON M·D·
Natus Aberdoniæ *Obiit Londini*
Anno 1620. *Anno 1683.*

Quæ Morisone viro potuit contingere major (*Ipse tibi patriam Phæbus concedit Apollo*,
Gloria, Pæonium quam superasse genus) (*Laureaque est capiti quælibet herba tuo .*
Archibaldi Pitcairne MD

Plantarum. The latter is essentially physiological, touching upon agriculture to a certain extent: the former is mainly morphological, structural, descriptive, and it is here that the first attempt at a classification of plants is to be found. In writing the *Historia*, Theophrastus was endeavouring, as a Greek philosopher rather than as a botanist, to "give account of" plants; and in order to do so he found it necessary to arrange them in some kind of order. Seizing upon obvious external features, he distinguished (*Lib.* I. *cap.* 5) and defined Tree, Shrub, Under-shrub and Herb, giving examples ; adding, however, that the definitions are to be accepted and understood as typical and general, "for some may seem perhaps to deviate" from them. Simple as was this mode of arrangement, Theophrastus further simplified it in the course of his work, by treating trees and shrubs as one group, and undershrubs and herbs as the other.

It may seem, at first sight, singular that a lecture purporting to discuss the state of systematic botany in England during the 17th century should begin with a reference to the botany of the Greeks. The explanation is that the elementary classification introduced by Theophrastus persisted throughout the 17th century ; the use of the groups Trees, Shrubs, and Herbs came to an end only in the 18th century, with the advent of Linnaeus. It seems almost incredible, but it is a fact, that the lapse of the nearly 2000 years that separated Theophrastus from Morison marked no material advance in the science of classification. Botanical works, when they were something more than com-mentaries on Theophrastus or Dioscorides, took cognizance of little else than the properties, medicinal or otherwise, of plants, and their economic uses.

A growing perception of the essential resemblances observable among plants can be traced, however, in the later Herbals, as they became less medical and economic and more definitely botanical. Thus, in the well-known work of Leonhard Fuchs (Fuchsius), *De Historia Stirpium Commentarii*, 1542, the plants are described in alphabetical order, without any reference to their mutual relation. But in Kyber's edition of Jerome Bock's (Tragus) *De Stirpium Nomenclatura, etc., Commentariorum Libri Tres*, published in 1552 (with a preface by Conrad Gesner), there is an attempt at a grouping of plants, though no principles are enunciated and no names are given to the groups, which resulted in the bringing together of labiate, leguminous, gramineous and umbelliferous herbs. The *Cruydtboeck* of Rembert Dodoens (Dodonaeus), 1554, marks much the same stage of progress, whereas the *Nova Stirpium Adversaria* of Pierre Pena and Matthias de l'Obel (Lobelius), issued in 1570, is a distinct step in advance. Here some idea is incidentally given of the principles that have been followed in the arrangement of the plants, but still no name is attached, as a rule, to the resulting groups. The work begins with an account of the herbaceous plants which, in modern terminology, are monocotyledonous : and at the end of the section (p. 65) de l'Obel thus explains what he has done :—" *Hactenus comparendo quot potuimus plantarum genera, quarum effigies et naturae ordinis consequutione ita sibi mutuo haererent, ut et facillime noscerentur et memoriae mandarentur, a Gramineis, Segetibus, Harundinibus, ad Acoros, Irides, Cyperos, hincque Asphodelos bulborum tuniceorum Caepaceorumve naturam praetervecti sumus.*" Cruciferous, caryophyllaceous, labiate and umbelliferous herbs are also segregated to some extent in the course of the work : and the leguminous herbs are brought together into a definite group, "*Alterum Frugum genus nempe graminis Trifolii et Leguminum,*" which is really the origin of the modern N. O. Leguminosae: though a few altogether foreign species, such as species of *Oxalis, Anemone Hepatica, Jasminum fruticans* L., and species of *Thalictrum*, are included among the trifoliate forms, and *Dictamnus Fraxinella* among the "Leguminosa." The *Stirpium Historiae Pemptades Sex sive Libri XXX* of Dodoens, published in 1583, shows considerable progress in classification as compared with his *Cruydtboeck* of 1554, more particularly in the recognition, apparently for the first time, of umbelliferous plants as a distinct group in a chapter headed *De Umbelliferis Herbis.*

Possibly these attempts to introduce some sort of system into Botany may have been inspired by the teachings of Conrad Gesner, that universal genius, who lived about this time (1516—1565). Though but fragments of his botanical writings have survived, it is clear from the much-quoted passage in a letter of his dated Nov. 26, 1565 (*Epistolae Medicae*, 1577, p. 113) that he too was seeking for the basis of a natural system of classification and that he thought he had found it in the flower and the fruit:—"*Ex his enim notis (a fructu, semine and flore) potius quam foliis, stirpium naturae et cognationes apparent.*"

Evidently at this period classification was in the air, and at length it began to precipitate and to crystallise in the work of Andrea Cesalpino (Caesalpinus: 1519—1603), Professor in the University of Pisa, whose *De Plantis Libri XVI*, published in 1583, is one of the most important landmarks in the history of systematic Botany. Here for the first time a system is propounded which is based definitely upon morphological observation. Cesalpino turns to the "fructification," that is the flower and the fruit, for his distinguishing characters. "*Enitamur igitur*," he says (*Lib.* I. *cap.* xiv.), "*ex propriis quae fructificationis gratia data sunt, plantarum genera investigare*"; and he goes on to point out that the observable differences here depend on number, position and form of the parts:—"*ad organorum constitutionem tria maxime faciant, scilicet, partium numerus, situs et figura.*" These principles he illustrates as follows:—the flower being the outermost covering of the fruit, a single flower may cover a single seed, as in the Almond: or a single seed-receptacle as in the Rose: or two seeds, as in the Umbelliferae: or two seed-receptacles, as in the Cress: or three seeds, as in the genus *Tithymalus* (*Euphorbia*); or three receptacles, as in the Bulbaceous plants (petaloid Monocotyledons): or four seeds, as in *Marrubium*: or four receptacles, as in *Euonymus*: or many seeds, as in the Cichoriaceae: or many receptacles, as in the Coniferae. The feature of the relative position of the parts which he especially emphasizes is whether the flower is inserted upon the top of the fruit (*i.e.* is epigynous): or is inserted lower around the fruit (hypogynous or perigynous). Moreover, the form of the seed, of the seed-receptacle, and of the flower, is to be taken into account.

The practical application of these principles led to a classification of plants which, though of course imperfect, was at least a good beginning. Following Theophrastus, Cesalpino divided plants into two main groups, (1) Trees and Shrubs, (2) Undershrubs and Herbs: each of these groups was then subdivided

according to the nature of the fruit and of the flower.   It will
be observed that Cesalpino, as was customary at that time,
designated as "seeds" all indehiscent one-seeded fruits, such as
nuts and the varieties of achene.   The following abstract will
suffice to give an adequate idea of the results obtained.   The
author's own words are given as nearly as possible.

ARBOREAE : **Seminibus saepius solitariis :**
Glandiferae : *e.g. Quercus.*
Vasculiferae : *Fagus, Castanea.*
Nuciferae : *Juglans, Carpinus, Corylus, Ulmus, Tilia, Acer,* &c.
Pericarpio tectae ; flore in sede fructus : *Prunus,* &c.
           flore in apice fructus : *Viburnum, Aesculus,* &c.
    Seminibus pluribus :
Flore carentes : *Ficus.*
Flos in summo fructus : *Morus, Sambucus, Hedera, Rosa,* &c.
Flos in sede fructus : *Vitis, Arbutus, Cornus,* &c.
Sedes seminis multiplex tecta communi corpore : *Pyrus, Citrus.*
   "   "   in siliquam producta : leguminous plants.
   "   "   bipartita : *Nerium, Syringa, Populus, Betula, Salix,* &c.
   "   "   tripartita : *Buxus, Myrtus.*
   "   "   quadripartita : *Vitex, Euonymus.*
   "   "   tecta proprio corpore : coniferous plants.
HERBACEAE : **Solitariis Seminibus :**
Semina nuda, papposa : *Valeriana.*
Semina pericarpio obducta : *Daphne, Jasminum.*
Flos in summo fructus : *Osyris, Valerianella.*
Flos in sede fructus, semen calyce exceptum : *Urtica,* Chenopodiaceae,
    Polygonaceae, Gramineae, Cyperaceae, Typhaceae.
    Solitariis Pericarpiis :
Flos exterius situs (Pomum) : Cucurbitaceae.
Flos inferius situs (Bacca) : Solanaceae, *Ruscus, Arum, Actaea,* &c.
    Solitariis Vasculis :
Legumina : leguminous herbs.
Capsulae : Caryophyllaceae, Primulaceae, Gentianaceae, &c.
    Binis Seminibus : (Genus Ferulaceum) Umbelliferae.
    Binis Conceptaculis :
Semina solitaria in singulis alveolis : Rubiaceae.
   "   plura, flore continuo : Scrophulariaceae, &c.
   "   "   flore in foliola quaterna diviso : Cruciferae.
    Triplici Principio, non-Bulbosae :
Semina nuda : *Thalictrum.*
   "   solitaria in tribus alveolis : Euphorbiaceae.
   "   plura in tribus alveolis : Convolvulaceae, Campanulaceae, &c.

Triplici Principio, Bulbosae:
Flos inferius sedet : bulbous Liliaceae.
Flos in summo fructus : Amaryllidaceae.
Bulbaceis ascribi desiderant : other Liliaceae, Iridaceae, Orchidaceae.
Quaternis Seminibus: Boraginaceae, Labiatae.
Pluribus Seminibus in communi sede: most Compositae.
Lactescentes : Cichorieae.
Acanaceae : Cynareae, *Dipsacus, Eryngium,* &c.
Pluribus Seminibus Flore communi :
Semina plene nuda : acheniferous Ranunculaceae and Rosaceae, &c.
Aut conjunctis receptaculis : *e.g. Aristolochia, Nymphaea, Papaver, Cistus.*
Aut disjunctis receptaculis : *e.g. Sedum, Veratrum, Helleborus, Delphinium, Dictamnus.*
Flore fructuque carentes : Cryptogams.

In spite of its inherent imperfections and of errors of observation, the method succeeded in bringing together a considerable number of the plants dealt with, into groups which are still regarded as natural. For instance, among the trees and shrubs, the leguminous genera, and the coniferous genera, respectively, are so brought together: and among herbs, the leguminous, umbelliferous, cruciferous and composite genera. Moreover, though many of Cesalpino's sections consist of what seems to be a heterogeneous assemblage of plants, yet they include groups of closely allied genera, representing several of the natural orders of more modern times, which his method was incapable of distinguishing. With all its shortcomings, the method produced a classification of plants which has proved to have been natural in no slight degree.

The very numerous botanical works which were published in the century after the appearance of Cesalpino's *De Plantis* afford evidence that his system of classification did not meet with an enthusiastic reception. Though his plant-names were generally quoted, his arrangement was entirely ignored : in fact the very idea of classification seems to have gradually faded out of the minds of botanists, whose attention was more and more engrossed with the description of the new species that the rapid extension of geographical discovery was bringing to light. This condition of the science is well illustrated by the most authoritative systematic work that the 17th century produced, the great *Pinax Theatri*

Botanici (1623) of Caspar Bauhin (1560—1624), a work which contains about six thousand plant-names, and was the product of forty years' labour. It might be expected that in such a work, special attention would have been paid to classification, that at least the best available system would have been used: as a matter of fact, the arrangement adopted is far inferior to that of Cesalpino and may be described as simply haphazard for the most part. The general lines of it are indicated by the following enumeration of the contents of the twelve Books of which the work consists ; the modern equivalents of his plant-names being given.

SUMMARY OF THE ARRANGEMENT ADOPTED IN BAUHIN'S *Pinax*.

Liber I.   Gramineae, Juncaceae, Cyperaceae, Typhaceae, *Ephedra*, *Equisetum*, *Hippuris*, *Asphodelus*, some Iridaceae, and Zingiberaceae.

Liber II.   De Bulbosis ; bulbous Monocotyledons, including Orchids with Orobanche, *Monotropa*, and *Lathraea*.

Liber III.   Olera et Oleracea ; most Cruciferae, Polygonaceae, and Chenopodiaceae, with some of the Compositae.

Liber IV.   Other Compositae ; *Delphinium*, *Fumaria* ; the Umbelliferae (so named) ; *Valeriana*.

Liber V.   Some Solanaceae, Papaveraceae, and Ranunculaceae ; *Gentiana*, *Plantago*, *Pyrola*, *Statice*, *Sarracenia*, *Nymphaea*, *Trapa*, *Sagittaria*, *Arum*, *Asarum*, and some Compositae.

Liber VI.   *Viola* ; *Cheiranthus*, *Matthiola*, *Alyssum*, *Hesperis* ; some Caryophyllaceae ; *Polygala*, *Specularia*, *Glaux*, *Linum*, *Cuscuta*, most Labiatae and Scrophulariaceae ; *Primula*, &c.

Liber VII.   *Lysimachia*, *Epilobium*, *Oenothera*, *Lythrum*, some more Labiatae, Scrophulariaceae, and Caryophyllaceae ; Boraginaceae ; some Compositae ; *Alisma* ; *Scabiosa* ; *Hypericum* ; Crassulaceae ; *Aloe* ; *Euphorbia*.

Liber VIII.   Various climbing plants ; *Convolvulus*, *Smilax*, *Humulus*, *Vitis* ; *Clematis*, *Lonicera*, *Hedera* ; and Cucurbitaceae : also Apocynaceae, Asclepiadaceae, some Liliaceae, Malvaceae, Rosaceae, Leguminosae, with other genera scattered among them, as *Aristolochia*, *Dentaria*, *Paeonia*, *Geranium*.

Liber IX.   Rubiaceae ; *Ruta*, *Thalictrum* ; the remainder of the Leguminosae.

Liber X.   Cryptogams in general : with a few scattered Phanerogams such as *Drosera*, *Oxalis sensitiva*, L. (*Herba viva foliis polypodii*) ; *Mimosa pudica* (*Herba Mimosa foliis Foenugraeci sylvestris*) ; *Lemna* ; and the remaining Compositae, the Thistles, with *Eryngium*, *Dipsacus*, and *Acanthus*.

Liber XI. Trees and Shrubs: Leguminous and Rosaceous; also *Rhus, Laurus, Fraxinus, Juglans, Castanea, Fagus, Quercus, Corylus, Tilia, Ulmus, Betula, Alnus, Populus, Acer, Platanus, Ricinus.*

Liber XII. *Mespilus, Crataegus, Berberis, Ribes, Sambucus, Ficus, Opuntia, Morus, Arbutus, Laurus, Daphne, Cistus, Myrtus, Vaccinium, Buxus, Olea, Salix, Ligustrum, Phillyrea, Rhamnus, Rubus Rosa, Tamarix, Erica,* Coniferous plants, *Palma.*

There was but one author, during this period, who made any material contribution to the science of classification, and that was Joachim Jung of Hamburg (1587—1657). Jung is best known by his *Isagoge Phytoscopica* (1678, ed. Vaget), the most philosophic and scientific treatise on plants that had appeared since the time of Aristotle, which is the foundation upon which the whole superstructure of plant-morphology and descriptive botany has since been erected. But it was in his *De Plantis Doxoscopiae Physicae Minores* (1662, ed. Fogel) that he expressed his views on systematic Botany. He did not propound a system of his own, but he sought to arrive at the principles upon which a classification should be based, with the logical result that he rejected the time-honoured Theophrastian division of plants into Trees and Herbs. Though Jung failed to produce any immediate impression upon the Botany of his time, he powerfully influenced the great developments which took place in the eighteenth century. It so happened that Ray, as he mentions in his *Index Plantarum Agri Cantabrigiensis* (1660), had obtained through Samuel Hartlib a MS. of the whole or part of Jung's *Isagoge*, which seems to have impressed him so much that he included many of Jung's morphological definitions in the glossary appended to the *Index*; and he subsequently embodied the *Isagoge* in the first volume of his *Historia Plantarum* (1686). It was from Ray's *Historia* that Linnaeus learned the morphological principles and terminology of Jung which were the basis of his own work in descriptive Botany, and rendered possible the elaboration of his system of classification. But, in spite of Jung, the venerable division of plants into Trees and Herbs continued to hold its own for a time. As will be seen, it was still adhered to by Morison and by Ray, even after it had been shown to be quite untenable by Rivinus (*Introductio Generalis in Rem Herbariam*) in 1690, and did not finally disappear until the time of Linnaeus.

It was just when systematic Botany had fallen back to its lowest level that Morison appeared upon the scene. He had been born at Aberdeen in 1620, and had there graduated Master of Arts with distinction by the time he was eighteen years old. His further studies in the natural sciences were interrupted by the Civil War, in which he took part on the Royalist side, being severely wounded in the battle of the Brig of Dee (1644). He fled to France, and there resumed his preparation for a scientific career with such success that he obtained, in 1648, the degree of Doctor of Medicine at the University of Angers. From that time onwards he devoted himself entirely to the study of Botany, which he pursued in Paris under the guidance of Vespasian Robin, Botanist to the King of France. In 1650 Morison was appointed by the Duke of Orleans, on Robin's recommendation, to take charge of the royal garden at Blois, a post which he held for ten years. The Duke of Orleans, shortly before his death early in 1660, had occasion to present Morison to his nephew King Charles II who was about to return to his kingdom. Soon after the Restoration, the King summoned Morison to London; and in spite of tempting offers made to induce him to remain in France, Morison obeyed the summons and was rewarded with the title of King's Physician and Professor of Botany with a stipend of two hundred pounds a year. During his tenure of these offices Morison found time to complete his first botanical work, the *Praeludia Botanica*, which was published in 1669; the same year in which he was appointed Professor of Botany in the University of Oxford.

A few words may be devoted, at this point, to the rise and progress of Botany in that University. In the year 1621, Lord Danvers (afterwards Earl of Danby), thinking " that his money could not be better laid out than to begin and finish a place whereby learning, especially the Faculty of Medicine, might be improved," decided to endow the University with a Physic Garden, such as was already possessed by various Universities on the Continent. With this object, he gave a sum of £250 to enable the University to purchase the lease of a plot of ground, about five acres in extent, situated "without the East Gate of Oxford, near the river Cherwell." A great deal of labour had to be expended upon the land after it had been secured : it was so low-lying that, as Anthony Wood says, "much soil was conveyed thither for the raising of the ground to prevent the overflowing of the waters" at the expense of Lord Danvers, who also caused to be built what Baskerville describes as " a most stately wall of hewen stone 14 foot high with 3 very considerable Gates thereto, one whereof was to the cost of at least five hundred pounds." The work proceeded but slowly, in consequence of the troublous times through which the country was passing, so that it was not completed until 1632. Even then the actual installation of the garden was delayed. About 1637 the Earl of Danby seems to have arranged with the well-known John Tradescant to act as gardener, but there is no evidence that Tradescant ever discharged the duties of the post: moreover, he died in the following year. Very shortly after this, though the exact date is not known, the Earl appointed Jacob Bobart to take charge of the Garden. Jacob Bobart was a German, born at Brunswick about the year 1599. He was an excellent gardener: under his care the garden flourished so well that the catalogue which was published in 1648 anonymously, though doubtless drawn up by Bobart, enumerated no less than 1600 species of plants in cultivation.

It had been the intention of Lord Danby to provide the University not only with a Physic Garden and a Gardener, but also with a Professor of Botany. For this purpose he bequeathed certain revenues : "but so it was that the times being unsettled, and the revenues falling short, nothing was done in order to the settling of a Professor till 1669." When the establishment of the Professorship had become possible, the University proceeded to elect Morison the first Professor of Botany, being influenced by the reputation which his recently published *Praeludia Botanica* had secured for him. Thus, after the lapse of nearly half a century, was Lord Danby's design completely realised.

Morison's chief occupation at Oxford was the preparation of his long promised *magnum opus*, the *Historia Plantarum Universalis Oxoniensis*. It was planned on a most extensive scale, and proved to be a laborious and costly undertaking. Morison impoverished himself in the preparation even of the one volume of it that appeared in his lifetime, though his many friends provided the cost of the 126 plates of figures with which it is illustrated, and the University advanced considerable sums of money. The work was to have been issued in three parts: the first part was to be devoted to Trees and Shrubs, and the other two parts to the Herbs. The volume published by Morison in 1680, and described as *Pars Secunda*, deals with only five out of the fifteen sections into which he classified herbaceous plants, although it extends to more than 600 folio pages. In the preface he gives as his reason for beginning with the Herbs rather than with the Trees and Shrubs, that he wished to accomplish first the most difficult part of his task lest, in the event of his death before the completion of the *Historia*, it should fall into the hands of incompetent persons. He did not live to finish his great undertaking. In November, 1683, he was in London on business connected with it: as he was crossing the Strand near Charing Cross, he was knocked down by a coach, and was so severely injured that he died on the following day. He was buried in the church of St Martin-in-the-Fields.

His unfinished work did not, as he feared, fall into incompetent hands. It was entrusted by the University to Jacob Bobart the younger, who on the death of his father in 1679, had succeeded him as Keeper of the Physic Garden, and who also succeeded Morison as *Horti Praefectus*, but not as *Professor Botanices*; the Professorship remained in abeyance for nearly forty years. After much difficulty and delay, a second and final instalment of the *Historia*, the *Pars Tertia*, dealing with the remaining ten sections of herbaceous plants, was published in 1699, as a folio of 657 pages with 168 plates. The material at Bobart's disposal was fairly abundant, consisting of Morison's MS. of four more of his sections of Herbs, with notes upon the remaining six sections. But even so, the task of completion was a laborious one, for it involved the incorporation of references to the very many descriptions of new plants that had been published since Morison's death: it has been generally admitted that Bobart discharged it with commendable skill.

Plate II

Great Gate of the Physic Garden, Oxford:
the elder Bobart in the foreground

The *Pars Prima*, that was to have been devoted to Trees and Shrubs, was never written. All that exists to represent it, is a stout MS. volume in the Library at the Botanic Garden, Oxford, apparently in Bobart's hand-writing, containing a classification and an enumeration of the species of trees and shrubs, which may possibly have been written with a view to publication.

A most interesting feature of Bobart's *Pars Tertia* is the *Vita Roberti Morisoni M.D.* with which the book opens, written by one of Morison's intimate friends, Dr Archibald Pitcairn. It is the source of all the available information regarding Morison up to the time of his coming to Oxford ; after that time much may be gathered concerning him from the records of the University. It is also a loyal defence of Morison and his system of classification against the criticisms to which, even then, he had been subjected. It concludes with a personal account of Morison, in which he is described as being " vigorous in body, having a mind trained to every kind of study, of ingenuous manners, calling a spade a spade, eager for true knowledge, a despiser of filthy lucre, considering the public advantage rather than his private gain." A portrait of him, here reproduced, forms the frontispiece to the volume.

Such was the life of the man whose botanical works are now to be considered : works that are not nearly so numerous as they are considerable, as will be seen from the following enumeration and brief description of them.

Praeludia Botanica, 1669 : a small 8vo volume of about 500 pages, which consists of the following parts :

(pp. 1—347) : *Hortus Regius Blesensis Auctus.*

(pp. 351—459) : *Hallucinationes Caspari Bauhini in Pinace, item Animadversiones in tres Tomos Universalis Historiae Johannis Bauhini.*

(pp. 463—499) : *Dialogus inter Socium Collegii Regii Gresham dicti et Botanographum Regium.*

Plantarum Umbelliferarum Distributio Nova, per Tabulas Cognationis et Affinitatis, ex Libro Naturae observata et detecta, 1672.

Plantarum Historiae Universalis Oxoniensis Pars Secunda,

seu Herbarum Distributio Nova per Tabulas Cognationis et Affinitatis ex Libro Naturae observata et detecta, 1680.

The three distinct treatises of which the *Praeludia Botanica* consists were written probably at different times, though published simultaneously in 1669. The first of them is an alphabetical catalogue, comprising about 2600 species, of the plants in the Royal garden at Blois when under Morison's care: 260 of the species are marked as new, and are fully described in an appendix. But the chief interest of the *Hortus Regius Blesensis Auctus* lies in the dedication to King Charles II. Morison here narrates how, whilst at Blois, he had framed a system of classification; how the King's Uncle, the Duke of Orleans, had promised to undertake the publication of a book to illustrate the system on an adequate scale, and how the sudden death of the Duke in 1660 had destroyed all such hopes; and he ends by appealing to the King to give him the patronage that he so much needed. " *Quod si annuere hoc mihi digneris,*" he wrote, "*polliceor Britanniam vestram cum methodo exactissima (quae est naturae ipsius) imposterum, in re Botanica gloriari posse, quemadmodum Italia, Gallia, Germania, superiori saeculo, sine methodo, in Scientia Botanica gloriatae sunt.*" But the King does not appear to have been moved by this dazzling promise. Morison evidently did not suffer from any lack of confidence in himself or in his method, of which he speaks on a previous page of the dedication, as " *methodus nova a natura data, a me solummodo (citra jactantiam) observata: a nullo nisi meipso in hunc usque diem detecta, quamvis mundi. incunabilis sit coeva,*" language which can hardly be described as modest. And yet, curiously enough, Morison gives not the slightest indication of the principles of this altogether new and original method of classification.

The second treatise, the *Hallucinationes*, is a searching and acute criticism of the published works of the brothers Bauhin: of the *Pinax* of Caspar, and of the *Historia* of John. Though he acknowledges in the preface the great value of their botanical labours, Morison did not fail to set out in detail the mistakes that they had made in both classification and nomenclature, and to make corrections which were, for the most part, justified.

Probably it was the critical study of the works of the Bauhins that led Morison to frame a system of classification of his own. The third and last treatise is the *Dialogus*: a dialogue between himself, as *Botanographus Regius*, King's Botanist, and a Fellow of the Royal Society, on the theme of classification. Here again Morison asserts the superiority of his own method : " *Methodum me observasse fateor : estque omnium quae unquam adhuc fuerunt exhibitae, praestantissima et certissima quippe a natura data.*" But he still fails to give any definite account of it : all that he says amounts merely to this, that the " *nota generica* " is not to be sought in the properties of a plant, nor in the shape of its leaves, as had been suggested by earlier writers, but in the fructification, that is, in the flower and fruit (*essentiam plantarum desumendam...a florum forma et seminum conformatione*).

The mention of a system of classification based on the form of the leaf evoked from *Botanographus* a pointed allusion to a book recently published by a Fellow of the Royal Society in which such a classification had been used, with the following severe comment : " *Ego tantum confusum Chaos : illic, de plantis legi, nec quicquam didici, ut monstrabo tibi et lapsus et confusionem, alias.*" The book so criticised was the encyclopaedic work edited by Dr John Wilkins, Bishop of Chester, and published by the Royal Society in 1668, entitled, " *An Essay towards a Real Character and a Philosophical Language,*" to which John Ray had contributed the botanical article ' *Tables of Plants.*' This criticism was the beginning of the unfriendly relations between Morison and Ray, of which some further account will be given subsequently.

Another point of interest in the *Dialogus* is the definite assertion (p. 488) that Ferns are ' perfect' plants, having flower and seed (*quia habent flores, qui fugiunt quasi obtutum, et semina quasi pulvisculum in dorso alarum*), an assertion which was repeated with even greater emphasis in Morison's preface to his edition of Boccone's *Icones et Descriptiones Rariorum Plantarum* etc. (Oxon. 1674), in opposition to the views of earlier writers, Cesalpino in particular. Cesalpino had, it is true, said of the group in which he had placed the Ferns and other Cryptogams, "*quod nullum semen molitur*" (*De Plantis*, p. 591): but he had added, in the same paragraph,—"*ferunt enim in folio quid, quod vicem seminis gerit, ut Filix et quae illi affinia sunt.*" It is a question if Morison was much nearer the truth than Cesalpino.

It is in the preface of his *Plantarum Umbelliferarum Distributio Nova* (1672) that Morison first gave a definite statement of the principles of his method, in the following terms: "*Cumque methodus sit omnis doctrinae anima: idcirco nos tam in hac umbelliferarum dispositione, quam in universali omnium stirpium digestione, quam pollicemur, notas genericas et essentiales a seminibus eorumque similitudine petitas, per tabulas cognationis et affinitatis disponentes stirpes exhibebimus. Differentias autem specificas a partibus ignobilioribus, scilicet radice, foliis et caulibus, odore, sapore, colore desumptas adscribemus, singulis generibus singulas accersendo species: ita species diversa facie cognoscibiles, sub generibus intermediis: genera intermedia sub supremis, notis suis essentialibus et semper eodem modo sese habentibus distincta militabunt. Hic est ordo a natura ipsa stirpibus ab initio datus, a me primo jam observatus.*"

It is not necessary to discuss in detail the merits of Morison's work on the Umbelliferae. It will suffice to say that it was published as a specimen of the great *Historia* that he had in preparation—*trigesimam operis quod intendimus partem*—so that the learned world might have some idea of what they were to expect from the completed work "*quemadmodum aiunt ex ungue leonem*"; and further, that it was the first monograph of a definite group of plants, and is remarkable for the sense of relationship between the genera that inspires it. The Umbelliferae constituted *Sectio IX* among the fifteen sections in which Morison distributed herbaceous plants.

At length, in 1680, appeared the *Pars Secunda* of the *Plantarum Historia Universalis Oxoniensis* in which work Morison's long-expected method of classification was to be exhibited and justified. However in this respect it proved to be disappointing: partly because it was so limited in its scope, dealing with but five of his fifteen *Sectiones* of herbaceous plants: and partly because it did not contain any complete outline of his system. It is most singular that, although he wrote so much, Morison should have died without having published any more definite information concerning his system of classification than what has been here cited.

Morison's influence did not, however, cease with his death; his tradition was maintained by the publication in 1699 of the *Pars Tertia* of the *Historia*, under the editorship of Bobart. This volume threw some welcome light upon Morison's system, inasmuch as it completed the description of the herbaceous plants, and gave a clear statement, in the form of a *Botanologiae Summarium*, of the classification resulting from the application of Morison's principles to these plants. But, even so, the revelation of the system still remained incomplete, in the absence of any account of the trees and shrubs.

It was not till nearly forty years after Morison's death, not until Bobart too was dead, that a full statement of Morison's method was published. In 1720 there appeared at Oxford a small tract of but twelve pages, the *Historiae Naturalis Sciagraphia*, containing an account of a complete system of classification, which agrees in all essentials, so far as herbaceous plants are concerned with that adopted by Morison and by Bobart in their respective volumes of the *Historia*: and, as regards trees and shrubs, with that in the MS. volume by Bobart which has been already mentioned. The tract is anonymous, but the matter that it contains is Bobart's work, whether it was written by himself or by some one who had access to his papers. This classification may be accepted as being essentially that of Morison, though somewhat modified by Bobart, who had undoubtedly been influenced by Ray's systematic writings which had appeared meanwhile. It is of such interest that it may be reproduced here, somewhat compressed, with an indication of the modern equivalents of the groups.

I. ARBORES.
    Coniferae semper virentes : most coniferous genera.
        „   foliis deciduis : *Larix, Alnus, Betula.*
    Glandiferae : *Quercus.*
    Nuciferae : *Juglans, Fagus, Corylus, Laurus,* &c.
    Pruniferae : *Prunus, Olea,* &c.
    Pomiferae : *Pyrus, Citrus, Punica, Ficus,* &c.
    Bacciferae : *Taxus, Juniperus, Morus, Arbutus, Sorbus,* &c

Siliquosae: *Cercis*, and other leguminous trees.
Fructu membranaceo: *Acer, Carpinus, Tilia, Fraxinus, Ulmus.*
Lanigerae non Juliferae: *Platanus, Gossypium.*
Juliferae et Lanigerae: *Populus, Salix.*
Sui generis Arbor: *Palma.*
II. FRUTICES.
Nuciferi: *Staphylea.*
Pruniferi: *Cornus.*
Bacciferi, foliis deciduis: *Viburnum, Rhus, Rosa, Ribes,* &c.
    „   semper virentes: *Ruscus, Phillyrea, Myrtus, Buxus,* &c.
Leguminosi: *Genista, Cytisus, Colutea.*
Binis Loculamentis: *Justicia, Syringa.*
Capsulis tetragonis: *Philadelphus, Tetragonia.*
    „   pentagonis: *Cistus.*
Multicapsulares: *Spiraea, Erica.*
Lanigeri: *Salix, Tamarix, Nerium.*
III. SUFFRUTICES.
Scandentes capreolis: *Vitis, Bignonia, Smilax.*
    „   viticulis: *Lonicera, Jasminum, Solanum,* &c.
    „   radiculis: *Hedera.*
IV. HERBAE.
Sectio i. **Scandentes**; Bacciferae: *Bryonia, Tamus,* &c.
                Pomiferae: most Cucurbitaceae.
                Campanulatae: Convolvulaceae.
Sectio ii. **Leguminosae, Papilionaceae siliquis bivalvibus**:
                Leguminous herbs.
Sectio iii. **Siliquosae Tetrapetalae Bicapsulares**: Cruciferae
                (with *Veronica* and *Polygala*).
                hisce adjiciuntur quaedam: *Chelidonium, Fumaria, Epi-
                lobium,* &c.
Sectio iv. **Hexapetalae Tricapsulares**:
Radicibus fusiformibus; *Asphodelus, Anthericum.*
    „   tuberosis; *Crocus, Gladiolus, Iris.*
    „   bulbosis; *Narcissus, Hyacinthus, Allium.*
    „   squamatis; *Lilium.*
Sectio v. **A Numero Capsularum et Petalorum Dictae**:
tricapsulares campanulatae; Campanulaceae.
    „   pentapetalae; *Hypericum, Viola.*
bicapsulares monopetalae; Scrophulariaceae.
quadricapsulares tetrapetalae; Rutaceae.
quinquecapsulares pentapetalae; Geraniaceae.
pentapetalae emollientes; Malvaceae.
    „   unicapsulares; Caryophyllaceae, Primulaceae.
    „   seminibus triangularibus; Polygonaceae.
    „   „   nigris splendentibus; Chenopodiaceae.

Sectio vi. **Corymbiferae** : (Compositae in part)
floribus aureis; *Artemisia, Tanacetum.*

„   rubris; *Adonis annua* L.

„   albis; *Bellis, Anthemis, Achillea,* &c.

„   ianthinis; *Xeranthemum, Scabiosa, Globularia.*

Sectio vii. **Flosculis Stellatis** : (the rest of the Compositae)
lactescentes non papposae; *Cichorium.*

„   papposae; *Lactuca, Sonchus, Hieracium.*

papposae non lactescentes; *Senecio, Aster, Doronicum,* &c.

„   capitatae; Cynareae.

Sectio viii. **Culmiferae seu Calamiferae** : Gramineae, Cyperaceae,
Typhaceae.

Sectio ix. **Umbelliferae.**

Hisce adnectuntur Plantae Stellatae; Rubiaceae.

Sectio x. **Tricoccae Purgatrices** : Euphorbiaceae.

Sectio xi. **Monopetalae Tetracarpae Galeatae et Verticillatae** :
Labiatae.

Hisce adjiciuntur Galeatae non verticillatae; *Verbena, Euphrasia.*
Et Verticillatae non Galeatae; *Urtica.*
Sequuntur Monopetalae tetracarpae asperifoliae; Boraginaceae.

Sectio xii. **Multisiliquae Polyspermae et Multicapsulares** :
multisiliquae; folliculate Ranunculaceae, *Sedum,* &c.

multicapsulares; *Papaver, Nymphaea,* Orchidaceae, *Aristolochia, Orobanche, Pyrola,* &c.

Sectio xiii. **Bacciferae** : some Solanaceae, *Sambucus, Cornus, Ruscus,
Arum,* &c.

Sectio xiv. **Capillares Epiphyllospermae** : Filices and Ophioglossaceae.

Sectio xv. **Heteroclitae seu Anomalae** : consists of

(*a*) Certain Phanerogams : *e.g. Piper, Acanthus, Apocynum,
Cuscuta, Reseda, Sagittaria, Alisma, Lemna, Drosera.*

(*b*) Pteridophyta other than Ferns : *Equisetum, Pilularia, Lyco-
podium.*

(*c*) Bryophyta, Algae, Fungi.

This then is the Morisonian method,—or at least the nearest
available approximation to it—in its entirety. The effect of its
application to the Vegetable Kingdom can hardly be accepted
as a sufficient justification of the superlatives with which its
author had introduced it. Of course it is not reasonable to judge
this method, or any other method of the past, by the standard
of botanical knowledge as at present existing : it can only be
fairly judged from the standpoint of its author. What has to
be considered is (1) the soundness of the principles adopted,

and (2) the consistency in the application of those principles. The conclusion to be drawn from such a consideration of the foregoing table is that Morison was more fortunate in his theory than in his practice. In spite of his statement that the "*nota generica*" should be taken from the fructification, many of the *Sectiones* are based upon quite other characters : such are (among the Herbs) the Scandentes, the Corymbiferae, the Culmiferae. Had Morison adhered more closely to his own principles, the results would have been more in accordance with his sanguine anticipations : such a heterogeneous group as *Sectio V*, for instance, would have been impossible. It was, perhaps, on account of its inconsistency that Morison's method never came into general use, although it was adopted enthusiastically by Paul Amman, Professor at Leipzig, in his *Character Plantarum Naturalis* (ed. 1685); and, with some modifications, by Christopher Knaut, Professor at Halle, in his *Enumeratio Plantarum circa Halam Saxonum sponte provenientium*, 1687, as well as by Paul Hermann, Professor at Leyden, in his *Florae Lugduno-Batavi Flores* (ed. Zumbach), 1690.

Morison's writings evoked severe contemporary criticism, more on account of their manner than of their matter. His constant reference to the "*Hallucinationes*" of Caspar Bauhin especially, was considered to be offensive even if warranted, for every botanist admitted a debt of gratitude to the author of the *Pinax*. Equally resented was Morison's oft-repeated statement that he had drawn the principles of his classification, not from the works of other writers, but from the book of Nature alone. It was urged against him that he had failed to do justice to his predecessors, particularly to Cesalpino : and it must be admitted that there is unfortunately some truth in this allegation. Morison's indebtedness to Cesalpino is suggested by the fact that the nature of the fruit, and in a secondary degree that of the flower, was the basis of both their methods. From a comparison of the two systems, as set out in this lecture, their fundamental resemblance can be traced through the many differences of detail. Since Morison does not quote Cesalpino in his books, it might be inferred that possibly he had not read him. But there is convincing evidence to the contrary. There is the fact that Morison's preface to the *Historia* contains a sentence taken *verbatim*, without acknowledgment, from the dedication of Cesalpino's *De Plantis*. Further, there is in the Library at the Oxford Botanic Garden a copy of the *De Plantis* containing many marginal notes which could not have been written by any one but Morison. The explanation of the position is probably this, that Morison regarded his classification as so great an advance upon that of Cesalpino, that he did not think it necessary to acknowledge what still remained of the earlier writer's work: but in any case his omission to mention Cesalpino was a grave error of judgment.

At this point it may well be asked, what are Morison's actual merits if, as it appears, he borrowed the leading principles of his classification from his predecessors? The most satisfactory answer to this question is that which is provided by those who lived and wrote at times but little removed from his own. Thus Tournefort, in his *Elemens de Botanique* (1694: p. 19) speaking of the work of Cesalpino and of Colonna, said—"*Peut-être que la chose seroit encore à faire si Morison...ne s'étoit avisé de renouveller cette metode. On ne sauroit assez louer cet auteur ; mais il semble qu'il se loue lui-même un peu trop : car bien loin de se contenter de la gloire d'avoir executé une partie du plus beau projet que l'on ait jamais fait en Botanique, il ose comparer ses découvertes à celles de Cristoffe Colomb, et sans parler de Gesner, de Cesalpin, ni de Columna, il assure en plusieurs endroits de ses ouvrages, qu'il n'a rien apris que de la nature même.*" Later, in his *Institutiones Rei Herbariae* (1700, p. 53) Tournefort expressed the same opinion in somewhat different words:—"*Legitima igitur constituendorum generum ratio Gesnero et Columnae tribui debet, eaque fortè in tenebris adhuc jaceret, nisi* Robertus Morisonus... *eam quasi ab Herbariis abalienatam renovasset, instaurasset, et primus ad usus quotidianos adjunxisset, qua in re summis laudibus excipiendus, longe vero majoribus si a suis abstinuisset.*"

The estimate formed of him by Linnaeus is clearly stated in a letter addressed to Haller probably about the year 1737: "Morison was vain, yet he cannot be sufficiently praised for having revived system which was half expiring. If you look through Tournefort's genera you will readily admit how much he owes to

Morison, full as much as the latter was indebted to Cesalpino, though Tournefort himself was a conscientious investigator. All that is good in Morison is taken from Cesalpino, from whose guidance he wanders in pursuit of natural affinities rather than of characters" (see Smith's *Correspondence of Linnaeus*, vol. ii. p. 281). If only Morison had frankly assumed the role of the restorer of a method that had been forgotten, instead of posing as its originator, his undoubted merits would have met with their just recognition, and his memory would have been free from any possible reproach.

Before Morison's method of classification could have come into general use, there was a rival system in the field, which was destined to achieve success, and in its course to absorb all that was good in Morison's: this was the system of John Ray.

Ray was born at Black Notley, near Braintree, Essex, on Nov. 29, 1628; so that he was not much junior to Morison. He studied and graduated with such distinction at the University of Cambridge, that he was in due course elected a Fellow of, and appointed a Lecturer in, his College (Trinity). Here he remained until 1662, when he resigned his Fellowship on his refusal to sign the declaration against 'the solemn league and covenant' prescribed by the Act of Uniformity of 1661. After leaving Cambridge he spent some years travelling both in Britain and on the Continent; and eventually settled at his birth-place, Black Notley, where he died on Jan. 17, 1704–5.

During his residence in Cambridge, Ray devoted much of his time to the study of natural history, a study which afterwards became his chief occupation. The first fruit of his labours in this direction was the *Catalogus Plantarum circa Cantabrigiam nascentium*, published in 1660, followed in due course by many works, for he was a prolific author, botanical and zoological as well as theological and literary, of which only those can be considered at present which contributed materially to the development of systematic botany.

The first such work of Ray's was his contribution of the *Tables of Plants* to Dr John Wilkins's *Real Character and a Philosophical Language*, published in 1669, which has already been mentioned in the course of this lecture (p. 21). The following

Plate III

Joannes Rajus
Societatis Regiæ Socius

is a summary of Ray's first attempt at a system of classification. He begins by distinguishing Herbs, Shrubs, and Trees. Proceeding to the detailed classification of Herbs, he divides them into *Imperfect* "which either do want or seem to want some of the more essential parts of Plants, viz. either Root, Stalk, or Seed," the Cryptogamia of Linnaeus; and *Perfect* "having all the essential parts belonging to a Plant." The Perfect Herbs are arranged in three main groups according to (1) their leaves, (2) their flowers, (3) their seed-vessel, each group being subdivided in various ways.

HERBS CONSIDERED ACCORDING TO THEIR LEAVES:

With long Leaves: Frumentaceous, "such whose seed is used by men for food, either Bread, Pudding, Broth, or Drink" (Cereals): or Non-Frumentaceous (other Grasses, Sedges, Reeds).

Gramineous Herbs of Bulbous Roots (Bulbous Monocotyledons).

Herbs of Affinity to Bulbous Roots (other Monocotyledons).

Herbs of Round Leaves (*e.g. Petasites, Viola, Pinguicula, Drosera*).

Herbs of Nervous Leaves (*e.g. Veratrum, Plantago, Gentiana, Polygonum*).

Succulent Herbs (*Sedum, Saxifraga*).

"Herbs considered according to the Superficies of their Leaves, or their Manner of Growing":

more rough (*e.g. Borago, Anchusa, Echium*):

less rough (*e.g. Pulmonaria, Symphytum, Heliotropium*):

stellate leaves (*e.g. Asparagus, Galium*).

HERBS CONSIDERED ACCORDING TO THEIR FLOWERS: "having no seed-vessel besides the Cup which covers the flower":

Herbs of Stamineous Flowers, "whose flower doth consist of thready Filaments or Stamina, having no leaves besides the *Perianthium*: or those herbaceous leaves encompassing these stamina, which do not wither or fall away before the seed is ripe"; and not of grassy leaves, may be distributed into such whose seeds are

Triangular (Polygonaceae);

Round: "distinguishable by sex, of male and female; because from the same seed some plants are produced which bear flowers and no seeds, and others which bear seeds and no flowers" (*e.g. Cannabis, Humulus, Mercurialis*): not distinguishable by sex (*e.g.* Chenopodiaceae, Urticaceæ, Resedaceae).

Herbs having a Compound Flower not Pappous } (Compositae).
Pappous Herbs   }

Umbelliferous Herbs (Umbelliferae, with *Valeriana*).

Verticillate Fruticose Herbs   } (Labiatae).
Verticillate Not Fruticose Herbs }

Spicate Herbs (a curious medley, including *Dipsacus, Eryngium, Echinops, Agrimonia, Circaea, Poterium Sanguisorba, Polygonum Persicaria, Trifolium stellatum, T. arvense*, and *Potamogeton angustifolium*).

Herbs bearing Many Seeds together in a Cluster or Button
(*e.g. Geum, Potentilla, Anemone, Ranunculus, Adonis, Malva*).

HERBS CONSIDERED ACCORDING TO THEIR SEED-VESSEL:

Of a divided Seed-vessel, which may be called Corniculate (*Paeonia, Dictamnus, Delphinium, Aquilegia, Aconitum, Geranium, Scandix*).

Of an entire Seed-vessel:

Siliquous: Papilionaceous Climbing Herbs } (Papilionaceae).
            Papilionaceous Herbs not Climbing

            Not papilionaceous (mostly Cruciferae).

Capsulate: bearing Flowers of Five Leaves (Caryophyllaceae, Hypericaceae, *Euphorbia, Linum, Lysimachia, Ruta, Nigella*).

            whose flowers consist of three or four Leaves (some Cruciferae, *Epimedium, Papaver, Verbena, Statice, Veronica*).

Campanulate Herbs:

            climbing (most Cucurbitaceae and Convolvulaceae):

            erect (Campanulaceae, some Solanaceae, *Digitalis*).

Not campanulate (Primulaceae, Scrophulariaceae, Acanthaceae, *Aristolochia, Vinca*).

Bacciferous herbs: may be distinguished according to their Qualities:

Esculent fruit: more pleasant (Strawberry),

                less pleasant (Tomato).

Esculent root (Potato):

Malignant: of simple leaves (Nightshade, Mandrake),

            of compound leaves (Herb Christopher, *Paris*).

Or Manner of Growth:

            being climbers (*Bryonia, Tamus, Smilax*):

            not climbers (*Physalis Alkekengi, Cucubalus, Sambucus Ebulus*).

OF SHRUBS.

I.   Bacciferous Spinous Shrubs of Deciduous Leaves
      (the genera *Rubus* and *Rosa*, Gooseberry, Sloe, Barberry, *Rhamnus, Lycium*).

II.   Bacciferous Shrubs of Deciduous Leaves, not Spinous
      (Vine, Currant, Bilberry, *Viburnum*, White Beam, *Cornus, Prunus Padus, P. Mahaleb, Diospyros*, Honeysuckle, Pepper, *Daphne, Euonymus*, Privet, *Salicornia*).

III. Bacciferous Sempervirent Shrubs
      (*Rhamnus Alaternus, Phillyraea, Arbutus, Daphne Laureola, Ruscus, Chamaerops humilis*, Laurustinus, Juniper, Myrtle, Ivy, Mistletoe).

IV.   Siliquous Shrubs
      (Lilac, *Cytisus, Colutea, Ulex, Genista, Mimosa*).

V. Graniferous Deciduous Shrubs
    (*Vitex, Spiraea, Tamarix, Jasminum, Althaea, Elaeagnus, Clematis, Ampelopsis*).
VI. Graniferous Evergreen Shrubs
    (*Cistus*, Oleander, Rosemary, *Phlomis fruticosa, Erica*, &c.).

OF TREES.

I.   Pomiferous Trees (Apple, Pear, &c., *Sorbus*, Fig, Pomegranate, Orange, Lemon, Banana).
II.   Pruniferous Trees (Peach, Plum, Cherry, &c., Olive, Date, Jujube).
III. Bacciferous Trees (Mulberry, Elder, Sumach, *Celtis*, Bay, Yew, Holly, Box, &c.).
IV.   Nuciferous Trees (Walnut, Almond, Hazel, *Castanea*, Beech, Coco-Palm, Coffee, Cocoa, Cotton).
V.   Glandiferous and Coniferous Trees (Oak, Alder, Larch, Cedar, Pine, Spruce, Cypress).
VI.   Trees bearing their Seeds in Single Teguments or Coverings (Carob, Tamarind, Elm, Hornbeam, Maple, Poplar, Willow. Lime, Plane).
VII. Trees considered according to their Woods or Barks (Lignum Vitae, Snakewood, Sandal-wood, Log-wood, Cinnamon, Cinchona, &c.).
VIII. Trees considered according to their Gumms or Rosins (Myrrh, Gum Arabic, Copal, Benzoin, *Liquidambar*, Camphor).

Such is the classification of which Morison spoke so slightingly in the *Dialogus*: though the character of the leaf is not made so much of as his criticism implied.   There is no need to dwell upon the strained relations that arose between Ray and Morison; it may suffice to say that Morison laid himself open to the charge of jealousy, and that Ray never forgave the criticisms, both written and oral, that Morison had made on him. Those who are interested in the unfortunate quarrel will find an account of it, with a most loyal apology for Morison, in Blair's *Botanical Essays* (1720).   Ray may certainly be acquitted of plagiarism which is suggested by Blair, for he had no opportunity of studying Morison's system in its entirety: since, as already explained, it was not published in a complete form until the appearance of the *Sciagraphia* in 1720, long after Ray's death.   When Ray wrote the *Tables of Plants* for Dr Wilkins, not even the *Preludia Botanica* had been published: the only work that he produced after the publication of both parts of Morison's *Historia* was the last edition of his *Methodus*

Plantarum (1703) which displays principles of classification of which Morison had no conception.

The *Tables of Plants* does not illustrate any very definite principles. It was a tentative production, written to order : in fact, it appears (as explained in the preface to his *Methodus emendata*, 1703) that Ray, in writing it, was not free to follow what he really believed to be the order of Nature. It is interesting, however, as being the first systematic work published in England. The classification is based, to some extent, upon the character of the fruit, a principle borrowed, probably not from Morison but directly from Cesalpino. Before long it was superseded by a much more comprehensive and ambitious attempt, the *Methodus Plantarum Nova*, issued in 1682, two years after Morison's *Historia* (*Pars Secunda*).

Ray's *Methodus Plantarum Nova*, 1682.

DE HERBIS.

Genus	i.	Imperfectae, flore et semine carentes : Algae, Fungi.
„	ii.	Semine minutissimo : Bryophyta, most Pteridophyta.
„	iii.	Acaules Epiphyllospermae, vulgo Capillares : Filices.
„	iv.	Flore imperfecto, sexu distinctae : *e.g. Humulus, Cannabis, Spinachia, Urtica.*
„	v.	„ imperfecto, sexu carentes : *e.g. Chenopodium, Alchemilla, Artemisia.*
„	vi.	„ imperfecto, Monospermae, semine triquetro : Polygonaceae.
„	vii.	„ composito, Lactescentes : Compositae, Cichorieae.
„	viii.	„ discoide, Papposae : Compositae, most Asteroideae and Senecionideae.
„	ix.	„ discoide nudo, Papposae : Compositae, *Eupatorium, Senecio, Gnaphalium.*
„	x.	„ composito discoide, Corymbiferae : Compositae, some Anthemideae.
„	xi.	„ discoide nudo, Corymbiferae : Compositae, the rest of the Anthemideae.
	xii.	„ ex flosculis fistularibus, Capitatae : Compositae, Cynareae.
„	xiii.	„ composito, Anomalae : *Dipsacus, Scabiosa, Echinops, Armeria.*
	xiv.	„ perfecto, seminibus nudis singulis : *Valeriana, Thalictrum, Statice, Agrimonia,* &c.
	xv, xvi.	Umbelliferae.

Genus   xvii.   Stellatae dictae: Rubiaceae.

„   xviii.   Asperifoliae: Boraginaceae.

„   xix, xx.   Verticillatae: Labiatae.

„   xxi, xxii.   Semine nudo, Polyspermae: acheniferous Ranunculaceae and Rosaceae, Malvaceae.

„   xxiii.   Pomiferae: Cucurbitaceae.

„   xxiv.   Bacciferae: *e.g. Smilax, Bryonia, Tamus*, some Solanaceae, &c.

„   xxv.   Multisiliquae seu Corniculatae: folliculate Ranunculaceae, *Sedum, Dictamnus*, &c.

„   xxvi.   ⎰Flore monopetalo uniformi: *e.g. Hyoscyamus, Gentiana,*
„   xxvii.   ⎱ *Convolvulus, Campanula.*
„   xxviii.   ⎱ Flore monopetalo difformi: *e.g. Impatiens, Aristolochia,* ⎰ most Scrophulariaceae.

„   xxix, xxx, xxxi.   Flore tetrapetalo uniformi siliquosae: Cruciferae.

„   xxxii.   Flore tetrapetalo uniformi, Anomalae: *e.g. Papaver, Ruta, Plantago, Veronica.*

„   xxxiii–vi.   Flore papilionaceo: Leguminosae.

„   xxxvii.   Flore pentapetalo aut polypetalo, foliis conjugatim dispositis: Caryophyllaceae, Cistaceae, Hypericaceae.

„   xxxviii.   Flore pentapetalo aut polypetalo, foliis nullo aut alterno ordine dispositis: *e.g. Portulaca, Viola, Reseda, Geranium, Linum.*

„   xxxix.   Flore pentapetaloide, Anomalae: *e.g. Primula, Asclepias, Erythraea, Verbascum.*

„   xl, xli.   Culmiferae: Gramineae.

„   xlii.   Graminifoliae non culmiferae: Cyperaceae, Juncaceae.

„   xliii–v.   Radice bulbosa: bulbous Monocotyledons.

„   xlvi.   Bulbosis Affines: *e.g. Iris, Aloe*, Orchidaceae, Araceae, *Cyclamen.*

„   xlvii.   Anomalae et sui generis: *e.g. Potamogeton, Nymphaea, Callitriche, Trapa, Stratiotes, Sagittaria, Cuscuta, Adoxa, Polygala.*

DE ARBORIBUS.

Genus   i.   Pomiferae: *Pyrus, Mespilus, Citrus.*

„   ii.   Pruniferae: *Prunus, Cornus, Olea, Palma.*

„   iii.   Bacciferae: *e.g. Myrtus, Laurus, Buxus, Arbutus, Ilex, Juniperus, Taxus.*

„   iv.   Nuciferae: *e.g. Juglans, Corylus, Quercus, Castanea, Fagus.*

„   v.   Coniferae: *Pinus, Cedrus, Abies, Cupressus, Larix, Betula, Alnus.*

„   vi.   Lanigerae: *Platanus, Tamarix, Salix, Populus.*

„   vii.   Siliquosae: leguminous trees, *Syringa.*

„   viii.   Vasculis seminum membranaceis et Anomalae: *Ulmus, Fraxinus, Carpinus, Tilia, Acer.*

O. B.

3

De Fruticibus.

Genus i.   Bacciferi sempervirentes: *e.g. Vaccinium, Ruscus, Hedera, Viscum, Juniperus.*

„   ii.   „   foliis deciduis, non spinosi: *e.g. Vitis, Lonicera, Cornus, Sambucus.*

„   iii. „   foliis deciduis, spinosi: *Crataegus* sp., *Ribes* sp., *Rosa, Berberis,* &c.

„   iv.   Seminibus nudis, aut vasculis siccis inclusis: *e.g. Vitex, Rhus, Spiraea, Erica.*

„   v.   Floribus papilionaceis: *e.g. Acacia, Genista, Cytisus, Colutea.*

„   vi.   Suffrutiscentes: a miscellaneous collection of species.

A comparison between the classification of the *Methodus Nova* and that of the *Tables of Plants* shows that whilst he left the Trees and the Shrubs almost unaltered, Ray remodelled his arrangement of the Herbs. Whereas, in the *Tables*, he had proceeded along three distinct lines of classification indicated by the characters of leaf, flower, and seed-vessel respectively, all regarded as equally important; in the *Methodus*, the leaf-character is subordinated to those of flower and fruit, and these are not kept distinct but are combined; a fundamental change of principle which is no doubt to be attributed to Morison's criticisms on the *Tables.* As Ray put it in his Preface: *Methodus haec differentias sumit a similitudine et convenientia partium praecipuarum, radicis puta, floris et ejus calicis, seminis ejusque conceptaculi.* The result is that many of the sub-divisions consist of groups of plants which are really natural, the precursors of several of the recognized Natural Orders of Phanerogams; such as Polygonaceae, Chenopodiaceae, Compositae, Umbelliferae, Rubiaceae, Boraginaceae, Labiatae, Cucurbitaceae, Scrophularia-ceae, Cruciferae, Leguminosae, Gramineae. The principles adopted were capable of yielding even better results, had they been more rigorously applied and had the investigation of the plants been more minute. For instance, in genera xxi and xxii, with a little more attention to floral characters, the Ranuncula-ceous might have been separated from the Rosaceous genera, and all of them from the Malvaceae: similarly in genera xxvi—xxviii, the Scrophulariaceous, and possibly also the Campanula-ceous genera, might have been segregated. One of the principal achievements is the recognition of the group Stellatae (Rubiaceae) as independent of, but related to, the Umbelliferae. For this, as well as other features, Ray was indebted to Cesalpino (conf. p. 11), as he acknowledges in his Preface. Nor does Ray fail to acknowledge his obligations to Joachim Jung, and to Morison whose *Preludia* and *Historia* he cites.

But if Ray's *Methodus Nova* owed something to Morison's *Historia* (*Pars secunda*), at a later stage the *Historia* (*Pars Tertia*) was even more indebted to the *Methodus Nova*. It is striking to observe how many of the groups constituted in the *Pars Tertia* and in the *Sciagraphia* (see p. 23) agree with those of Ray. It is this close association, amounting almost to mutual dependence, of the systems of these two botanists, that makes comparative criticism of them an impossibility. Their relative position may, in fact, be summed up in the statement that both of them adopted the principles of Cesalpino, and that Ray eventually proved to be more successful than Morison in their application.

The *Methodus Nova* is something more than a system of classification. The systematic part of the work is preceded by five *Sectiones* which are morphological essays bearing the following titles: I. *De Plantarum seminibus observationes quaedam generales*: II. *De Foliis Plantarum seminalibus dictis*: III. *De Plantula seminali reliquisque semine contentis*: IV. *De Floribus Plantarum, eorumque partibus et differentiis*: V. *De Divisione Plantarum generali in Arbores, Frutices, Suffrutices et Herbas*. Beginning with the last, it is a discussion of the propriety of retaining the old Theophrastian sub-divisions: Ray agreed with Jung (see p. 15) that they are popular rather than accurate and philosophical, but he retained them on the ground of expediency. The fourth *Sectio* is an outline of the morphology of the flower based upon Jung's *Isagoge* which Ray had received in MS. from Dr John Worthington who had obtained it from Samuel Hartlib, as is explained in the Preface. The first three *Sectiones* are of peculiar interest: they give an account of Ray's observations upon seeds and seedlings, with quotations from Malpighi's recent work on the same subject (*Anatomes Plantarum, Pars Prima*, 1675; *Pars altera*, 1679), recognizing the fact that the seedlings of some plants have two seed-leaves or cotyledons (as Malpighi first called them), those of others only one, a fact which came to be of great systematic importance. The classification of the *Methodus Nova* was maintained by Ray in his *Historia Plantarum* (t. i, 1686), as well as in both the first (1690) and second (1696) editions of his *Synopsis Methodica Stirpium Britannicarum*, somewhat improved and more compact in form. His ultimate views were expressed in the *Methodus Plantarum emendata et aucta*, published in 1703 not long before his death. In many respects this final form of his system is a great improvement upon that of 1682; more especially in the adoption of the number of the seed-leaves as a systematic character. Ray, it is true, limited the application of this character to herbaceous plants, as he had not brought himself to give up the old categories of Herbs, Shrubs and Trees nevertheless, he founded in this work the groups of *Dicotyledones* and *Monocotyledones* which persist, though materially altered as to their content, to the present day.

Ray's *Methodus Emendata et Aucta*, 1703.

    DE HERBIS. **Flore Destitutae.**

    Genus i. Submarinae: Algae, &c.

      „ ii. Fungi.

      „ iii. Musci: Bryophyta with *Lycopodium*.

      „ iv. Capillares: Filices.

          Herbae sui generis: *Ophioglossum, Pilularia, Salvinia, Salicornia*, &c.

          Floriferae. *Dicotyledones.*

      „ v. Flore stamineo: *e.g.* Urticaceae, Polygonaceae, Chenopodiaceae, &c.

      „ vi–ix. Flore Composito seu aggregato: Compositae, with Dipsaceae, *Eryngium, Globularia.*

      „ x. Flore simplici, semine nudo solitario: *e.g. Valeriana, Mirabilis, Agrimonia.*

      „ xi. Umbelliferae.

      „ xii. Stellatae: Rubiaceae.

      „ xiii. Asperifoliae: Boraginaceae.

      „ xiv. Verticillatae: Labiatae.

      „ xv. Semine nudo, Polyspermae: *e.g. Alisma, Ranunculus, Potentilla.*

      „ xvi. Pomiferae: Cucurbitaceae.

      „ xvii. Bacciferae: *Bryonia, Tamus, Arum, Polygonatum, Solanum*, &c.

Genus xviii. Multisiliquae: folliculate plants, *e.g. Delphinium, Asclepias, Sedum.*

„   xix. Vasculiferae Flore monopetalo: (capsulate Gamopetalae).
Regulari; Campanulaceae, Primulaceae, Malvaceae, Gentianaceae, &c.
Irregulari; Scrophulariaceae, *Aristolochia, Acanthus*, &c.

„   xx. Tetrapetalae Siliquosae et Siliculosae: Cruciferae.
Anomalae; *Papaver, Euphorbia, Epilobium*, &c.

„   xxi. Flore Papilionaceo, sive Leguminosae.

„   xxii. Pentapetalae Enangiospermae sive Vasculiferae: (capsulate Polypetalae), *e.g.* Caryophyllaceae, Cistaceae, Hypericaceae, Geraniaceae, Violaceae.

Monocotyledones.

Genus xxiii. Graminifoliae Tricapsulares, radice bulbosa, tuberosa, fibrosa:
Flore fructus basi cohaerente; Liliaceae.
Flore summo fructui insidente; Iridaceae, Amaryllidaceae.
Bulbosis Affines: *Cyclamen*, Orchidaceae, Zingiberaceae.

„   xxiv. Graminifoliae Flore stamineo; Gramineae, Cyperaceae, Typhaceae.

„   xxv. Anomalae aut Incertae Sedis: *e.g. Nymphaea, Trapa, Epimedium, Sarracenia, Piper*, &c.

DE ARBORIBUS ET FRUTICIBUS: A. **Flore a Fructu remoto**: (diclinous or dioecious plants).

Genus i. Coniferae: *Abies, Pinus, Cedrus, Cupressus, Larix, Betula, Alnus.*

„   ii. Non-Coniferae:
Floribus racematim dispositis stamineis: *Buxus, Pistacia.*
„ in fasciculos congestis: *Empetrum.*
Juliferae: nuciferae: *Juglans, Corylus, Carpinus, Quercus, Fagus.*
piluliferae: *Platanus.*
lanigerae: *Populus, Salix.*
Bacciferae: *Juniperus, Taxus, Morus.*

B. **Flore Fructui contiguo**:

Genus i. Umbilicatae; flore summo fructui insidente:
Pomiferae: *Pyrus, Sorbus, Rosa, Punica*, &c.
Bacciferae, Polypyrenae: *Ribes, Sambucus, Hedera*, &c.
„ Monopyrenae: *Viburnum, Cornus*, &c.

„   ii. Non-Umbilicatae; flore basi fructus cohaerente:
Pruniferae: *Prunus, Olea.*

Pomiferae: *Citrus.*

Bacciferae, Monopyrenae: *Viscum, Daphne, Rhamnus* sp.

    „   Polypyrenae: *e.g. Vitis, Rubus, Ligustrum, Berberis,* &c.

Genus iii. Fructu sicco, non Siliquosae: *e.g. Acer, Fraxinus, Tilia, Ulmus, Rhus, Syringa.*

    „   iv. Siliquosae Flore non papilionaceo: *Cassia, Mimosa, Ceratonia, Nerium,* &c.

    „   v. Siliquosae Flore papilionaceo: papilionaceous plants.

    „   vi. Anomalae: *Ficus.*

Foliis Arundinaceis: Monocotyledons; Palmaceae, *Dracaena, Bambusa.*

There can be no doubt that Ray was more fortunate than Morison in the impression that he produced upon contemporary botanists and upon those who immediately succeeded them. This, for instance, is what Tournefort said of him (*Elemens de Botanique*, 1694, p. 19): "*Monsieur Ray sans faire tant de bruit a beaucoup mieux réussi que Morison. Sa modestie est louable, et l'Histoire des Plantes qu'il nous a donnée est une Bibliotheque Botanique, dans laquelle on trouve non seulement tout ce que les auteurs ont dit de meilleur sur chaque plante; mais encore les caracteres des genres y sont designez d'une maniere assez commode....*" In the *Classes Plantarum* (1738) Linnaeus gave a somewhat formal approval of Ray's work: "*Magna sunt opera J. Raji in Scientia Botanica, qui constantia summa, omnia, quae beneficio seculi innotuerant de plantis, manu plus quam ferrea descripsit.*" But perhaps a more genuine opinion is that expressed by Linnaeus in the letter to Haller from which his estimate of Morison has already been quoted (see p. 27): "You are here justly aware, that when the System of Ray was spoken of as perfectly natural, all botanists must have been blind, unless, like Dillenius, they hoped for a professorship, or were compelled, by the authority of the English, to give to Ray supreme honours. What was he? Undoubtedly an indefatigable man in collecting, describing, etc.; but in the knowledge of generic principles, less than nothing, and altogether deficient in the examination of flowers. I beg of you to compare the first edition of his *Methodus* with the second and third, where he has learned to take everything from Tournefort. I know not why the discoveries of Caesalpinus have escaped all observation, whilst everything has stupidly been ascribed to Ray" (Smith's *Correspondence of Linnaeus*, ii. p. 280—1). This rather severe criticism does not, however, seem to have prejudiced Haller against Ray, for in the former's well-known *Bibliotheca Botanica* (vol. i. p. 500, 1771), in speaking of the rapid progress of Botany in the latter part of the seventeenth century, he adds—"*Multa pars horum incrementorum debetur Johanni Ray. Vir pius et modestus, V. D. M. maximus ab hominum memoria botanicus, ea felicitate usus est, ut totos quinquaginta annos dilecto studio ei licuerit impendere.*"

Ray's system also became more popular than that of Morison, and was in general use in England until the latter half of the eighteenth century, when it was gradually superseded by the Linnean method which was first applied to English botany in Dr J. Hill's *Flora Britannica* (1760).

Ray was never engaged in teaching any branch of natural history. Had there been, in his day, a Chair of Botany in the University of Cambridge, he would, no doubt, have occupied it: however, the professorship was not established until 1724, twenty years after his death. He might very well have been chosen to succeed Morison at Oxford: but, for some unstated reason, the professorship there was kept in abeyance for nearly forty years after the death of Morison.

As has been explained, Morison and Ray revived the forgotten labours of Cesalpino. The immediate result of the publication of their systems was to stimulate their colleagues on the continent of Europe to a noble emulation: there was scarcely a botanist of note who did not elaborate a system of his own. After suffering from too little work in the direction of classification, botany now began to suffer from too much: one after the other, system followed system in rapid succession. Those, for instance, of Christopher Knaut (1687), Paul Hermann (1690), Boerhaave (1710), Rivinus (1690—1711), Ruppius (1718), Christian Knaut (1716); and, in France, of Tournefort (1694, 1700), and of Magnol (1720). Then came the *Methodus Sexualis* of Linnaeus (*Systema Naturae*, 1735). The effect of the general adoption of Linnaeus' most useful but artificial method was the temporary arrest almost everywhere, except in France, of the quest of the natural system. Though this was the effect of the introduction of his method, it was not at all the intention of Linnaeus: for in his *Classes Plantarum* (1738, p. 485) he said, "*Primum et ultimum in parte Systematica Botanices quaesitum est Methodus Naturalis.*" On the same page of that work he laid down, in a series of aphorisms, the principles upon which alone the construction of such a method can be successfully attempted; and he gave special emphasis to this one, that the classificatory characters should not be taken from a single structure but from all: "*nec una vel altera pars fructificationis, sed solum simplex symmetria omnium partium.*" It was just because they had failed to formulate this principle that the earlier systematists,—whether Fructists, as Cesalpino, Morison, Ray, Knaut and Hermann; or Corollists, as Rivinus and Tournefort; or Calycists, as Magnol—were not more successful, and that their systems, even the *Methodus emendata* of Ray, were more or less artificial.

It was in France that the carving out, as it were, of the Natural Orders from the solid block of genera was carried on with the greatest success. This process had become much less difficult since Tournefort had begun to constitute genera in the modern sense of the term. Before his time the word "genus" had been applied indiscriminately to every kind of plant-group (see the systems of Cesalpino and Ray, pp. 12, 32): the largest groups were the *summa genera*; the smaller, the *genera subalterna* or *infima*. Tournefort limited the application of the term to the smallest groups of species, designating by the term *Classe* the largest groups which he subdivided into *Sections* (*Elemens de Botanique*, 1694). It was Linnaeus (*Classes Plantarum*, p. 485) who introduced the term *Ordo* to designate the subordinate groups of the classes.

Tournefort himself succeeded, by means of his corollist method, in distinguishing for the first time the following *Sections*, describing their flowers by terms which are now familiar as the names of natural orders; *Flore Labiato, Cruciformi, Rosaceo, Caryophyllaceo, Liliaceo, Papilionaceo, Amentaceo*; though these sections do not all exactly agree with the modern

Natural Orders of similar designation.   A remarkable, if not altogether successful, attempt in the same direction was Adanson's *Familles des Plantes* (1763), based upon the sound Linnean principle, " *qu'il ne peut i avoir de Methode naturele en Botanicke, que celle qui considere l'ensemble de toutes les parties des Plantes.*" The number of species and varieties known in his day amounted to something over eighteen thousand: these, reduced into 1615 genera, he grouped into fifty-eight families.   Several of those had been already more or less well defined; but most of them were entirely original, and not a few of them persist to the present day, though Adanson is not credited with all that are his due.   His lack of method in naming his families, to say nothing of the fantastic nomenclature of his genera, made it necessary for other names to be preferred to his.   Still some familiar names of natural orders are attributable to him, such as *Hepaticae, Onagrae, Compositae, Caprifolia, Borragines, Portulacae, Amaranthi, Papavera, Cisti,* though most of them have since undergone some change in their termination.   In addition to these, there are several which would have been credited to Adanson, had it not so happened that they had also been suggested by Bernard de Jussieu: such are, *Palmae, Aristolochiae, Myrti, Campanulae, Apocyna, Verbenae, Thymeleae, Gerania, Malvae, Ranunculi.*   Adanson was the first to publish these names (1763): but Bernard de Jussieu had made use of them as early as 1759 in laying out the Trianon Garden at Versailles, though they were not actually published until 1789, when all the 65 orders devised by him were included in the *Genera Plantarum secundum Ordines Naturales disposita* of his famous nephew Antoine Laurent de Jussieu.   Here at last was a fairly complete natural system, consisting of one hundred natural orders arranged in fifteen classes, within the three great subdivisions, *Acotyledones, Monocotyledones, Dicotyledones,* constituting the framework of that which is accepted at the present day.   It has undergone many modifications, of which the first and most important were those effected by A. P. de Candolle (*Théorie Élémentaire*, 1813), who, while he improved upon Jussieu in various ways, made the unfortunate, but happily unsuccessful, attempt to substitute "*Endogenae*" for "*Monocotyledones*" and

" *Exogenae* " for " *Dicotyledones*." The system has proved itself capable of expansion to accommodate all the new genera and natural orders that have since been established: it has justified itself as a natural classification in its susceptibility to development in precision as well as in extent, and in that it has survived the many experiments made upon it during the first century of its existence.

The glory of this crowning achievement belongs to Jussieu: he was the capable man who appeared precisely at the psychological moment, and it is the men that so appear who have made, and will continue to make, all the great generalisations of science. Jussieu's achievement, like other great scientific achievements, would have been impossible without the labours and failures of his predecessors, of which some account has been given in this lecture. He himself attributed much of his success to the work of Tournefort, but it is clear that he owed at least as much to Ray: if he learned from the former the systematic importance of the gamopetalous and of the polypetalous corolla, he gleaned from the latter the value of the cotyledonary characters upon which are based his three primary subdivisions of the Vegetable Kingdom.

It has been necessary to go beyond the strict limits of the history of British Botany in order to make it clear to what extent and at what period our two distinguished fellow-countrymen contributed to the development of the natural system of classification. Enough has been said to establish the importance and the opportuneness of their contributions: if Pisa was glorified by the birth of Systematic Botany, and Paris by its adolescence, Oxford and Cambridge were honoured by its renascence. The question concerning the respective merits of Morison and Ray finds perhaps its most satisfactory answer in the words of Linnaeus (*Classes Plantarum*, 1747, p. 65):—"*Quamprimum Morisonus artis fundamentum restaurasset, eidem mox suam superstruxit methodum Rajus, quam dein toties reparavit, usque dum in ultima senectute emendatam et auctam emitteret*": Morison relaid the foundation upon which Ray built. As Linnaeus points out, Ray enjoyed the advantage of a very long period of productive activity: in the thirty-four years that separated his

Tables of Plants from his *Methodus Emendata et Aucta*, he had time to revise and remodel his system. Morison, on the contrary, was prevented by unfavourable circumstances from beginning the publication of his Method until late in life, and he was not permitted to see more than a fragment of it issue from the press.

It is probable that Ray was more truly a naturalist than was Morison: for in addition to his works on Method, he published not only his *Catalogus Plantarum circa Cantabrigiam nascentium* (1660), but also a *Catalogus* of British plants (1670, 2nd ed. 1677), almost the earliest work of the kind, only preceded by William How's *Phytologia Britannica* (1650), which developed into the first British Flora arranged systematically, the *Synopsis Methodica Stirpium Britannicarum* (1690, 2nd ed. 1696). Morison published nothing on field-botany; his volume of the *Historia* contains, it is true, occasional mention of plants found in or near Oxford, but the finder of them seems always to have been the younger Bobart. Ray included in the *Synopsis* a list of plants that had been communicated to him by Bobart, with whom he seems to have been intimate, and expressed his indebtedness to Bobart's botanical skill.

But whether the palm be bestowed upon the one or the other, the fact remains that both were men of exceptional capacity, and that both did good work for British Botany, raising it to a level which commanded the respect and admiration of the botanical world; from which, as the succeeding lectures of this course will show, it was not allowed to sink. What Linnaeus said of Morison may be applied equally to Ray,—"*Roma certe non uno die, nec ab uno condebatur viro. Ille tamen faces extinctas incendit, a quibus ignem mutuati sunt subsequentes, quibus datum ad lucidum magis focum objecta rimare*" (*Classes Plantarum*, p. 33).

NEHEMIAH GREW

1641—1712

By AGNES ARBER

Ancestry and Life narrative—his versatility—state of Botany—Grew and Malpighi—Grew's *bona fides* vindicated—*The Anatomy of Vegetables Begun*—seed structure—his treatise on the Root—its dedication—*The Anatomy of Trunks*—*The Anatomy of Plants*—illustrations—Grew's conception of cells and tissues—the plant as a textile fabric—analogy with the animal body—medullary rays—secondary thickening—his understanding of external morphology—physiological notions—suggestions for experiments—importance of the habitat—the sexes of flowers—floral and seed structure—estimate of his contributions to Botany.

NEHEMIAH GREW, who, with the Italian botanist Marcello Malpighi, may be considered as co-founder of the science of Plant Anatomy, lived in stirring and troubled times. His life[1] extended from 1641 to 1712; that is to say, he was born the year before King Charles I proclaimed war upon the parliamentary forces, and he lived through the Protectorate, the reigns of Charles II, James II, William and Mary, and the greater part of the reign of Queen Anne. He came of a stock remarkable for courage and independence of mind. His grandfather, Francis Grew, is described as having been a layman, originally of good estate, but "crush'd" by prosecutions for non-conformity in the High Commission Court and Star Chamber. Francis Grew had a son Obadiah, who was a student of Balliol, and entered the Church. When the Civil War broke out, he sided with the parliamentary party, but was by no means a blind adherent of Cromwell, with whom he is said to have pleaded earnestly for the life of King Charles I. In 1662 Obadiah Grew resigned

[1] *Dict. Nat. Biog.*, edited by Leslie Stephen and Sidney Lee, vol. XXIII. 1890.

Plate IV

NEHEMIAH GREW (1701)

Portrait of Nehemiah Grew after the portrait by R. White which is
reproduced in the *Cosmologia Sacra*, 1701

his living, being unable to comply with the Act of Uniformity. Twenty years later, as a man of seventy-five, he was convicted of a breach of the Five Mile Act, and imprisoned for six months in Coventry Gaol. But though by this time his sight had failed, his spirit was indomitable. Whilst in prison, he dictated a sermon every week to an amanuensis, who read it to several shorthand writers, each of whom undertook a number of copies; it was then distributed to various secret religious meetings, at which it was read. Nehemiah Grew was Obadiah's only son, and it is a curious fact that the year 1682, which witnessed the father's imprisonment, was the year in which the son published his *magnum opus, The Anatomy of Plants,* prefaced by an Epistle Dedicatory to "His most sacred Majesty Charles II." So far as one can gather, Nehemiah Grew's career seems to have been singularly unaffected by the political crises that took place around him. The deliberate style of his writing certainly suggests a studious and unruffled life. He was an undergraduate at Pembroke Hall, Cambridge, and afterwards took his doctor's degree in medicine at Leyden, at the age of thirty. He seems to have been successful in his profession, and we learn from the sermon[1] preached at his funeral that he died suddenly, whilst still actively engaged in his practice. In the words of the sermon, "It was his Honour and Happiness, to be Serviceable to the last Moments of Life."

Before turning to Grew's botanical work, it may be worth while to refer very briefly to his writings on other subjects, showing as they do the remarkable versatility of his mind. He produced a series of chemical papers, and also pamphlets on the method of making sea-water fresh, and on the nature of the salts present in the Epsom wells. In 1681 appeared his *Musæum Regalis Societatis,* a catalogue raisonné of the objects in the Museum of the Royal Society, with which were bound up some contributions to animal anatomy. The Catalogue is a bulky volume, and it is hard to forbear a smile on reading that Grew dedicated it to one Colwall, the founder of the Museum, in order

---

[1] Enoch's Translation. A Funeral Sermon Upon the Sudden Death of Dr Nehemiah Grew, Fellow of the College of Physicians. Who died March 25th, 1712. Preach'd at Old-Jewry. By John Shower. London. 1712.

that the Royal Society "might always wear this Catalogue, as the Miniature of [his] abundant Respects, near their Hearts." As we should expect, this Catalogue is far more discursive than such a work would be if it were drawn up at the present day, though Grew takes credit to himself for not "medling with Mystick, Mythologick, or Hieroglyphick matters." He manages, however, to introduce some general remarks which are of interest. He realises, for instance, that it is possible to group living creatures in a way which has some significance, and that it is the business of the biologist to discover this grouping. He blames Aldrovandus for beginning his history of quadrupeds with the horse, because it is the most useful animal to man, and points out that Gesner's arrangement, which is purely alphabetical, is even less satisfactory. "The very Scale of the Creatures," he concludes, "is a matter of high speculation." It is tempting to quote largely from the Catalogue, but I will confine myself to one other remark of Grew's which is perhaps particularly applicable to-day, when the quotation of authorities is apt to become almost an obsession: "I have made the Quotations," he says, "not to prove things well known, to be true ;...as if *Aristotle* must be brought to prove a Man hath ten Toes."

Grew's last work was the *Cosmologia Sacra*[1], a folio volume occupied with a defence of Christianity, and an explanation of the author's views on the nature of the Universe. There is a copy in the British Museum, the earlier part of which is crowded with marginal and fly-leaf notes, in some cases initialled or even signed in full by Samuel Taylor Coleridge. One cannot help recalling Charles Lamb's humorous complaint that books lent to Coleridge were apt to be returned "with usury; enriched with annotations tripling their value...in *matter* oftentimes, and almost in *quantity* not unfrequently, vying with the originals." Coleridge seems to have accepted Grew quite seriously as a thinker. In one of his manuscript notes we read, "It is from admiration of Dr N. Grew, and my high estimate of his Powers, that I am almost tempted to say, that the Reasonings in Chapt. III *ought* to have led him to the perception of the *essential phænomenality* of Matter." That these reasonings did *not* so lead him, must,

[1] 1701.

I think, be attributed to the fact that Grew was above all things
a naturalist, and Coleridge a philosopher, and that between the
two an intellectual gulf is often fixed.

After this somewhat lengthy introduction, it is more than
time to turn to our main subject,—the study of Nehemiah
Grew's work as a botanist.

Botanical science was in a decidedly decadent condition
when Grew entered the field. The era of the herbal was closing.
The last English book of any importance which can strictly be
included under this head, Parkinson's *Theatrum Botanicum*, was
published the year before Grew was born, and a lull in this
kind of work followed. It is true that Culpeper's *Herbal*
appeared later, but this bombastic work was of no botanical
value. It was reserved for Morison and Ray to open a new
era in British Systematic Botany. At the same time, fresh
inspiration was being breathed into the science from quite a
different quarter. The herbalists studied plants primarily with
a view to understanding their medicinal properties. Nehemiah
Grew also approached Botany in the first instance from the
medical standpoint, but it was his knowledge of anatomy which
opened his mind to the possibility of similar work, with the
bodies of plants, instead of those of animals, as the subject.
He tells us that he was impressed by the fact that the study
of animal anatomy had been carried on actively from early
ages, whereas that of vegetable anatomy had been scarcely so
much as contemplated. "But considering," he continues, "that
both came at first out of the same Hand, and are therefore the
Contrivances of the same Wisdom; I thence fully assured my
self, that it could not be a vain Design, though possibly unsuccessful, to seek it in both."

Grew was drawn to the study of plant structure at the age of
twenty-three, and seven years later he produced his earliest work
on the subject, *The Anatomy of Vegetables Begun*, which was
published by the Royal Society in 1672. It will be remembered
that the Royal Society was then quite in its youth, its first
beginnings only dating back to about 1645[1]. By a curious
coincidence,—recalling the classic case of Darwin and Wallace

[1] *Life of Robert Boyle* by Thomas Birch, p. 83, 1744.

at the Linnean Society,—on the very day that Grew presented
his treatise in print, the Secretary of the Royal Society received
Marcello Malpighi's manuscript dealing with the same subject.
Priority can however be fairly claimed for the Englishman, since
he had submitted his treatise to the Society in manuscript earlier
in the year. This question of priority, and also the question
whether Grew was guilty of plagiarism from Malpighi's writings,
has been much discussed at different times. Schleiden[1] in particular brought forward charges of the most serious nature
against Nehemiah Grew's good faith. These accusations were,
however, dealt with in detail in a pamphlet by Pollender[2] in
1868, and shown to be groundless,—Schleiden's information
about the circumstances being wholly inaccurate. There is
now practically no doubt that Grew was an independent worker,
and was only definitely indebted to Malpighi, in so far as he
himself acknowledges it. In the preface to the second treatise,
for instance, he mentions the Italian botanist, and remarks in
speaking of the "Air-vessels"—"the manner of their Spiral
Conformation (not observable but by a Microscope) I first
learned from Him, who hath given a very elegant Description
of them." If Grew had been a wholesale plunderer from
Malpighi's writings, he would scarcely have been likely to have
acknowledged indebtedness on a special point. It must be confessed, however, that judging by present-day standards of scientific
etiquette, Grew should have referred more fully to the works[3] of
the Italian author, in his final book, *The Anatomy of Plants*.

The Anatomy of Vegetables Begun contains more that is of
interest from a morphological than from a strictly anatomical

---

[1] M. J. Schleiden, *Grundzüge der wissenschaftlichen Botanik*, Vol. I. p. 198,
1842. The incorrect statement that Grew was Secretary of the Royal Society at the
time that Malpighi's manuscript was received by that body, is also repeated in the
English translation of Schleiden's work [Schleiden's *Principles of Scientific Botany*,
translated by Edwin Lankester, London, 1849, p. 38].

[2] Aloys Pollender, *Wem gebührt die Priorität in der Anatomie der Pflanzen dem
Grew oder dem Malpighi?* Bonn, 1868.

[3] Marcellus Malpighi, *Anatome Plantarum*, 2 pts, London, 1875 and 1879 (see
also Marcellus Malpighi, *Die Anatomie der Pflanzen, Bearbeitet von M. Möbius*,
Leipzig, 1901. In this little book the more important parts of Malpighi's work are
translated into German, and a number of the figures reproduced).

Plate V

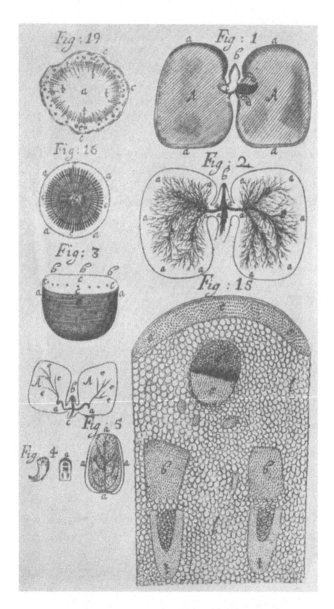

Plate from *Anatomy of Vegetables Begun*, 1672

Figs. 1—4, Bean Seed; 1, Bean opened out; 2, Same to
shew 'seminal root'; 3, 'Lobe' cut across; 4, 'Plume'
cut across. Fig. 5, Gourd and Lupine Seeds. Figs. 15,
16, 19, Anatomy of Burdock

standpoint, according to the modern sense of the terms. In botanical language, the meaning of the word anatomy has become restricted since Grew's time, until it is now often used to denote microscopic detail alone. Grew devotes a good deal of space to the study of seed structure, dealing chiefly with such features as can be observed with the naked eye (Pl. 5). He invented the term "radicle" for the embryonic root, and used the word "plume" for the organ which we now speak of in the diminutive as the plumule. The cotyledons he called "lobes," but he recognised that they might in some cases appear above ground and turn green, becoming in his terminology "dissimilar leaves." He took the Bean seed as his principal type, and described it with the lucid picturesqueness which is so characteristic of his writing. It is, he says[1], "cloathed with a double Vest or Coat: These Coats, while the Bean is yet green are separable and easily distinguished. When 'tis dry, they cleave so closely together, that the Eye, not before instructed, will judge them but one; the inner Coat likewise (which is of the most rare contexture) so far shrinking up, as to seem only the roughness of the outer, somewhat resembling Wafers under *Maquaroons*. At the thicker end of the Bean, in the outer Coat, a very small *Foramen* presents it self:...That this *Foramen* is truly permeable even in old setting *Beans*, appears upon their being soak'd for some time in Water: For then taking them out, and crushing them a little, many small Bubbles will alternately rise and break upon it."...The Plume "is not, like the *Radicle*, an entire Body, but divided at its loose end into divers pieces, all very close set together, as Feathers in a Bunch; for which reason it may be called the *Plume*. They are so close, that only two or three of the outmost are at first seen: but upon a nice and curious separation of these, the more interiour still may be discovered.... In a French Bean the two outmost are very fair and elegant. In the great Garden-Bean, two extraordinary small Plumes, often, if not always, stand one on either side the great one now describ'd." These two "extraordinary small plumes" are, in other words, the structures which we should now describe as buds in the axils of the cotyledons. Grew also notices that two

[1] The order of the paragraphs cited is slightly altered from that of the original.

simplified leaves are borne next above the cotyledons, or, as he
expresses it, the "Plume" is "cooped up betwixt a pair of
Surfoyls."

Grew deals also with the vernation of leaves, and methods of
bud protection. He shews that their position and folding gives
"two general advantages to the Leaves, Elegance and Security,
sc. in taking up, so far as their Forms will bear, the least room;
and in being so conveniently couch'd, as to be capable of receiving
protection from other parts, or of giving it to one another; as
for instance, First, There is the *Plain-Lap*, where the Leaves
are all laid somewhat convexly one over another, but not plaited;
being to the length, breadth and number of Leaves most agreeable; as in the Buds of *Pear-tree*, *Plum-tree*, etc. But where
the Leaves are not thick set, as to stand in the *Plain-lap*, there
we have the *Plicature*; as in *Rose-tree*, *Strawberry*, *Cinquefoyl*,
Burnet etc." Grew refers also to rolled vernation, distinguishing
between the "Fore-Rowl" and the "Back-Rowl." He thus
remarks on the hairy covering characteristic of young leaves :—
"the *Hairs* being then in form of a *Down*, always very thick set,
thus give that protection to the Leaves, which their exceeding
tenderness then requires; so that they seem to be vested with
a Coat of *Frieze*, or to be kept warm like young and dainty
Chickens, in Wooll."

In the year following the publication of *The Anatomy of
Vegetables Begun*, Nehemiah Grew produced a second treatise,
under the title, "An Idea of a Phytological History Propounded.
Together with a Continuation of the Anatomy of Vegetables;
Particularly prosecuted upon Roots. And an Account of the
Vegetation of Roots Grounded chiefly thereupon." In the
dedications of his books Grew often reveals much of his own
personality, and of his attitude towards science, although such
revelations are apt to be mingled with the curious "conceits,"
and extravagant flattery, characteristic of the time. For instance
he dedicated this particular work to the President and Fellows
of the Royal Society, and after addressing to them some apolo-
getic remarks about his own performance, he takes heart of
grace from the thought that "how unpromising soever the Stock
may be, yet the Fruit cannot but be somewhat matured upon

which You are pleas'd to shine." It shews how strong the influence of fashion can be, when we find such bombast coming from the pen of a man who, only a few lines earlier, has written, with the perfection of simplicity, "Withal, I looked upon Nature as a Treasure so infinitely full, that as all men together cannot exhaust it; so no man, but may find out somewhat therein, if he be resolved to try."

The most important part of this treatise is the account of the comparative structure of roots, to which we will return later, when discussing Grew's anatomical conceptions. With regard to the position of the plant in the soil, he held somewhat mystical views. He believed that the "air-vessels," or tracheal elements, tended to draw the plant upwards, and the roots to pull it downwards. He says, for example, that the upper part of the roots of most seedlings ascend, because the first leaves being large and standing in the open air, "the *Air-vessels* in them have a dominion over the young *Root*, and so yielding themselves to the sollicitation of the *Air upwards*, draw the *Root* in part after them."

In 1675 appeared Grew's third botanical work, *The Comparative Anatomy of Trunks*, which dealt with stem structure, as the previous work dealt with root structure. There is, in the British Museum, a particularly interesting copy of this book, which is elaborately annotated in manuscript. From internal evidence it seems almost certain that this is the author's copy, corrected in his own handwriting[1]. Some, though not by any means all, of the corrections are identical with the alterations found in the 1682 edition. Above the first plate is written "vide ye Book Interleavd," and we may perhaps hazard the guess that in this copy we have Grew's first suggestions, whilst those which he finally adopted in the second edition were inserted in the interleaved copy whose whereabouts, if it still exists, is unknown at the present time.

Pl. 6 shews a typical page from the annotated copy. At

---

[1] By the courtesy of the Council of the Royal Society, I have been able to compare these annotations with certain manuscript letters of Nehemiah Grew's preserved in the Society's Library. This comparison confirms the view that the annotations are in Grew's own handwriting.

the foot we find the note "Air-Vessels out of Parenchyma, transformed, as Caterpillars to Flys," shewing that Grew had arrived at some idea of the formation of vessels. The whole section of the book to which this page belongs is very much remodelled in the 1682 edition, but the analogy just quoted is introduced and Grew proceeds accurately to describe the origin of vessels. "And as the *Pith* it self, by the Rupture and Shrinking up of several *Rows* of *Bladders*, doth oftentimes become Tubulary: So is it also probable, that in the other *Parenchymous Parts*, one single *Row* or *File* of *Bladders* evenly and perpendicularly piled ; may sometimes, by the shrinking up of their Horizontal *Fibres*, all regularly breakone [sic] into another and so make one *continued Cavity.*"

I have passed over these three treatises in a somewhat cursory fashion, because Nehemiah Grew's botanical work is perhaps better studied in his final pronouncement on the subject,—a folio volume published in 1682 under the title of "The Anatomy of Plants. With an Idea of a Philosophical History of Plants. And several other Lectures, Read before the Royal Society." This work consists of second editions of his three earlier treatises, largely rewritten, with a great deal of additional matter, including a section on the anatomy of flowers, and many new figures. Some of the plates are excellent, and especially remarkable for the way in which Grew shews the anatomy in drawings which represent the organ in three dimensions (Pl. 7). He himself laid great stress on this. In his own words, " In the *Plates*, for the clearer conception of the *Part* described, I have represented it, generally, as entire, as its being magnified to some good degree, would bear....So, for instance, not the *Barque*, *Wood*, or *Pith* of a *Root* or *Tree*, by it self; but at least, some portion of all three together : Whereby, both their *Texture*, and also their Relation one to another, and the *Fabrick* of the whole, may be observed at one *View.*" One cannot help wishing that botanists of the present day would more often take the trouble to illustrate their papers on this principle.

It is as a plant anatomist that Grew is chiefly famous, and it is important to try to realise exactly how far his conception of the anatomical structure of plants has been confirmed by

Plate VI

The Comparative ... Ch. 3.
omy of *Roots*. ... for our clearer understanding the *Texture* of the said *Veſſels*, I
will a little further illuſtrate the ſame by
this ſimilitude. I would reſemble it thus.
As if a ... ſhould be wound ſpirally, and
edg, to edg, round about a Stick; and ſo
the Stick being drawn out, the ...
ſhould be left, in the Figure of a *Tube*, an-
ſwerable to an *Air-veſſel*. ...
For that which upon the *unroveing* of the
Veſſel, ſeems to be a *Plate*; is really *Natural*
... : being *not one intire piece*; ...
... but
conſiſting of a certain Number of *Round Fi-*
bers, ſtanding collaterally, as the *Threds*
do in a ... And are alſo,
much after the ſame manner, by other
Croſs and ſmaller *Fibers*, ſtitched cloſe up
together; as is moſt apparent in the large
and elder *Air-veſſels* of *Vine*, *Oak*, and many other *Plants*. What theſe *Croſs Fibers*
are, will better be underſtood when we
come to the *Texture* of the *Bith*.

CHAP.

A page from *The Comparative Anatomy of Trunks*, Nehemiah Grew,
1675. The annotations are believed to be in the author's own
handwriting. [British Museum. Printed Books Dept. (972. *a.* 10)]

more recent research. In appraising his work it must be remembered that he was essentially the pioneer of the science. It is true that some observations on plant anatomy occur in Robert Hooke's *Micrographia*, which was published six years before Grew sent in his first manuscript to the Royal Society ; but Hooke never really attempted to make a systematic study of the subject. He had succeeded in greatly improving the microscope, and his chief interest was in applying his instrument to all kinds of bodies, vegetable and otherwise. Cork, charcoal, pith, etc., came under his observation, and to some extent he understood their structure. Grew acknowledges indebtedness to "the Learned and most Ingenious Naturalist Mr *Hook*," and tells us that some of the results which Hooke obtained, inspired him to study certain of his plants again with a better microscope. For instance Hooke was able to see smaller pores in wood than Grew had been able to detect, but, with better glasses, he confirmed the accuracy of Hooke's observation. However, although Hooke must certainly be credited with priority in the discovery of the fact that plant tissues are characterised by a cellular structure, his botanical work, considered in its entirety, is of very slight significance compared with that of Grew.

Grew's clearest account of plant cells is perhaps to be found in his description of root parenchyma, which he compares to "the Froth of *Beer* or *Eggs*" or to "a fine piece of *Manchet*[1]," or again, to "a most curious and exquisitely fine-wrought Sponge." He quotes with approval Hooke's description of *Elder-Pith* as "an heap of *Bubbles*." It would be unsafe however to conclude that he had really arrived at what is known as the Cell Theory. His conception of the nature of plant tissues was not by any means that of the modern botanist. He believed the cell-walls to consist of interwoven fibres, which were continuous from cell to cell. He did not consider that these fibres were invariably wrought together in such a fashion as to enclose bladder-like spaces, or cells; in some cases he held that the tissue was non-cellular, consisting simply of interwoven fibres. It was these hypothetical fibres, rather than the cells, which he

[1] Manchet=a loaf of fine wheaten bread. (*An Etymological Dictionary of the English Language*. W. W. Skeat. New ed. 1910.)

regarded as of fundamental importance. His idea, which is somewhat confusing, is perhaps best understood from his comparison of plant structure with pillow lace. The "most unfeigned and proper resemblance we can," he writes, " at present make of the whole *Body* of a *Plant*, is, To a piece of *fine Bone-Lace*, when the Women are working it upon the *Cushion*, For the *Pith*, *Insertions*[1], and *Parenchyma* of the *Barque*, are all extream Fine and Perfect *Lace-Work*: the *Fibres* of the *Pith* running *Horizontally*, as do the *Threds* in a Piece of *Lace*; and bounding the several *Bladders* of the *Pith* and *Barque*, as the *Threds* do the several *Holes* of the *Lace*; and making up the *Insertions* without *Bladders*, or with very small ones, as the same *Threds* likewise do the *close* Parts of the *Lace*, which they call the *Cloth-Work*. And lastly, both the *Lignous* and *Aer-Vessels*, stand all *Perpendicular*, and so cross to the *Horizontal Fibres* of all the said *Parenchymous Parts*; even as in a Piece of *Lace* upon the *Cushion*, the *Pins* do to the *Threds*. The *Pins* being also conceived to be *Tubular*, and prolonged to any length ; and the same *Lace-Work* to be wrought many Thousands of times over and over again, to any thickness or hight, according to the hight of any *Plant*. And this is the true *Texture* of a *Plant*."

Grew thus visualised the inner structure of the plant as a textile fabric, and the analogy between vegetable substance and woven threads seems to have been constantly present in his mind. The same idea also occurs, for instance, in the dedication of his *magnum opus*, where he says, "one who walks about with the meanest *Stick*, holds a Piece of Nature's Handicraft, which far surpasses the most elaborate *Woof* or *Needle-Work* in the World."

The notions at which Nehemiah Grew arrived on the subject of the vascular anatomy of plants were more advanced than his ideas on the ultimate nature of the tissues. There is no doubt that the comparison with animal anatomy, which was constantly in his mind, was on the whole helpful, though it led to some errors. The following paragraph, which occurs in the *Cosmologia Sacra*, seems to be an instance in which the analogy with the

---

[1] Medullary rays.

animal kingdom, helped him to take a broad view. " In the
Woody Parts of Plants, which are their Bones; the Principles
are so compounded, as to make them Flexible without Joynts,
and also Elastick. That so their Roots may yield to Stones,
and their Trunks to the Wind, or other force, with a power of
Restitution. Whereas the Bones of Animals, being joynted, are
made Inflexible."

In plants, as in animals, Grew looked for " vessels," and discovered by means of a simple experiment that continuous tubes,
worthy of being called by this name, existed in the outer parts
of the root, whereas the pith consisted of closed chambers. He
cut a fresh root transversely, and then gently pressed the side of
it with his finger nail. He was able to detect the vessels with the
naked eye, and he observed that where they occurred, sap oozed
out under pressure, but was sucked in again when the pressure
was removed. The pressure also expressed a certain amount
of sap from the pith, where vessels were absent, but here the sap
was not sucked in again when the root was no longer squeezed,
shewing that the liquid had only been forced out by the wounding of the cells. Had they been open tubes like the vessels, the
release of the pressure would have caused the sap to disappear.
Grew recognised that the vascular tissue of the root is centrally
placed, whereas in the stem it is circumferential, and he points
out that this difference is connected with the diverse mechanical
needs of the two organs. It should also be noted that he discovered that concentration of the vascular system is characteristic
of climbing plants, the wood, in his own words, standing " more
close and round together in or near the Center, thereby making
a round, and slender *Trunk*. To the end, it may be more
tractable, to the power of the external *Motor*, what ever that
may be : and also more secure from breaking by its winding
Motion." He observed the radial arrangement of the xylem in
the root, and offered an explanation of it, which is however
scarcely free from obscurity. " Some of the more Æthereal
and Subtile parts of the *Aer*, as they stream through the *Root*,
it should seem, by a certain *Magnetisme*, do gradually dispose
the *Aer-Vessels*, where there are any store of them, into *Rays*."
Amongst other details of root anatomy, Grew discovered that

all the tissues outside the central cylinder sometimes peel off when the root becomes old, or as he says, "the whole body of the Perpendicular *Roots*, except the woody *Fibre* in the Centre, becomes the second *skin.*" Turning to stem structure, we find that he understood the difference in origin between stem buds and adventitious roots. The stem bud, he writes, "carries along with it, some portion of every *Part* in the *Trunk* or *Stalk*; whereof it is a *Compendium.*" The adventitious root, on the other hand, "always shoots forth, by making a Rupture in the *Barque*, which it leaves behind, and proceeds only from the inner part of the *Stalk.*" He describes the vascular bundles of the stem as "fibres" perforated by numerous "pores." It would be a mistake, however, to suppose that he had no understanding of their structure, at least as regards the xylem, for he goes on to say that "each *Fibre*, though it seem to the bare eye to be but *one*, yet is, indeed, a great number of *Fibres* together; and every *Pore*, being not meerly a space betwixt the several parts of the Wood, but the *Concave* of a *Fiber.*" He noticed the medullary rays, for which he uses the expressive term "Insertions." "These *Insertions*," he says, "are likewise very conspicuous in Sawing of *Trees* length-ways into Boards, and those plain'd, and wrought into *Leaves* for *Tables*, *Wainscot*, *Trenchers*, and the like. In all which,...there are many parts which have a greater smoothness than the rest; and are so many *inserted Pieces* of the *Cortical Body*; which being by those of the *Lignous*, frequently intercepted, seem to be discontinuous, although in the *Trunk* they are really extended, in continued Plates, throughout its Breadth."

Nehemiah Grew was interested in the process of secondary thickening, but he only arrived at a dim notion of how it took place. He grasped, however, the important point that in a tree trunk the meristematic zone lies near the surface, "the young *Vessels* and *Parenchymous Parts*" being formed annually "betwixt the *Wood* and *Barque.*" He describes how, "every year, the *Barque* of a *Tree* is divided into Two Parts, and distributed two *contrary* ways. The outer Part falleth off towards the *Skin*; and at length becomes the *Skin* it self....The inmost portion of the *Barque*, is annually distributed and added to the *Wood*;

Plate VII

From Grew's *Anatomy*

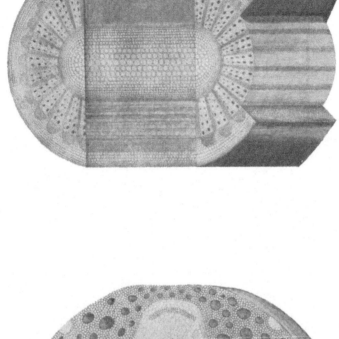

Part of a Vine Branch cut transversly, and
splitt half way downe yͤ midle

Sheweth the Parts of a Goosberry

the *Parenchymous Part* thereof making a new addition to the *Insertions* within the *Wood*; and the *Lymphæducts* a new addition to the *Lignous pieces* betwixt which the *Insertions* stand. So that a *Ring* of *Lymphæducts* in the *Barque* this year, will be a *Ring* of *Wood* the next; and so another *Ring* of *Lymphæducts*, and of *Wood*, successively, from year to year." Exactly what Grew meant by the term "Lymphæduct" is not always clear. In some cases he seems to refer to the phloem and cambium by this name, and in other cases to the perimedullary zone. The annual rings in Oak, Elm, Ash, etc. came under his observation, and he remarks that the difference between the Spring and Autumn wood, as we should now call it, arises from the fact that "the *Aer-Vessels* that stand in the inner *margin* of each annual *Ring*, are all vastly bigger, than any of those that stand in the outer part of the *Ring*."

Grew did not enter into the minuter details of histology, except in his description of the spiral tracheids, to which, as we have seen, his attention was first called by Malpighi's observations. He speaks of the spiral as formed of "Two or More round and true *Fibres*, although standing collaterally together, yet perfectly distinct. Neither are these Single *Fibres* themselves *flat*, like a *Zone*; but of a *round* forme, like a most fine *Thred*." He makes the curious statement that the direction of the spiral is constant, being "in the *Root*, by *South*, from *West* to *East*: but in the *Trunk*, contrarily, by *South*, from *East* to *West*."

Although it is as an anatomist that Nehemiah Grew is best known, his grasp of external morphology is perhaps even more remarkable. His work on seed structure has already been quoted. He seems to have quite readily detected the true nature of modified stems. He examined for instance the thorns of the Hawthorn, and saw that their structure was axial. In his own words, they "are constituted of all the same substantial *Parts* whereof the *Germen* or *Bud* it self [is], and in a like proportion: which also in their Infancy are set with the resemblances of divers minute *Leaves*." It should be recalled that Albertus Magnus, the great scholastic philosopher, writing in the thirteenth century, distinguished between thorns and prickles, and noticed

transitions between the former and leafy branches[1]. There is
no reason to suppose, however, that our author was acquainted
with the work of Albertus. Grew realised the nature of Bulbs,
and points out that "the *Strings* only, are absolute *Roots* ;
the *Bulb*, actually containing those *Parts*, which springing
up, make the *Leaves* or *Body*; and is, as it were, a Great *Bud*
under ground."

Nehemiah Grew was interested in plant physiology, although
the state of chemical and physical knowledge at the time did
not allow of his advancing so far in this, as in the morphological
side of the subject. His turn of mind, too, appears to have
naturally led him to the study of form rather than that of function.
As regards the absorption of water, his idea was simply that
the roots sucked up water like a sponge, because the parenchyma
was of a spongy nature. He supposed that the liquid was
rendered purer by being strained through the skin, which, according to whether it was of a texture resembling brown paper,
cotton, or leather, would produce a different effect upon any
solution passing through it. His explanation of the ascent of
the sap had really much in common with the "Kletterbewegung"
theory propounded by Westermaier[2] almost exactly two hundred
years later. Grew argued that "considering to what height and
plenty, the *Sap* sometimes ascends ; it is not intelligible, how
it should thus ascend, by virtue of any one *Part* of a *Plant*,
alone ; that is neither by virtue of the *Parenchyma*, nor by virtue
of the *Vessels*, alone." He pointed out that the parenchyma
might suck up a liquid for a short distance, and also the vessels,
like "small *Glass-Pipes* immersed in Water, will give it an
ascent for some Inches ; yet there is a certain *period*, according to the *bore* of the *Pipe*, beyond which it will not rise." To
account for the rise he supposes that the vessels and parenchyma
work together, the turgidity of the surrounding parenchyma
cells both compressing the vessels, and thus causing the liquid
in them to ascend, and also actually forcing some of their own
contents into them.

[1] Ernst H. F. Meyer, *Geschichte der Botanik*, vol. IV. p. 60, 1857.
[2] M. Westermaier, "Zur Kenntniss der osmotischen Leistungen des lebenden
Parenchym's." *Ber. d. deutsch. bot. Gesellsch.* Bd 1. p. 371, 1883.

Grew performed a few experiments, especially in the direction of plant chemistry. This was a natural line of work for a doctor, since the extraction of various vegetable substances had long been practised in medicine. He noticed, amongst other points, that the green infusion obtained by treating a plant with olive oil would, at least in the case of certain aromatic plants, appear of a green colour in a small drop, but of a red, or deep yellow, when a quantity of it was held up against a candle. In other words, Grew seems to have observed the characteristic fluorescence of chlorophyll.

He was interested also in the subject of geotropism, and succeeded in proving that there is an innate tendency for the root to grow down and the stem to grow up; and that it is not merely a case of the root seeking the soil, and the stem the air. His directions for performing the experiment are as follows :—" Take a Box of *Moulds*, with a hole bored in the bottom, wide enough to admit the *Stalk* of a *Plant*, and set it upon stilts half a yard or more above ground. Then lodg in the *Mould* some *Plant*, for Example a *Bean*, in such sort, that the *Root* of the *Bean* standing in the *Moulds* may poynt upwards, the *Stalk* towards the ground. As the *Plant* grows, it will follow, that at length the *Stalk* will rise upward, and the *Root* on the contrary, arch it self downward. Which evidently shews, That it is not sufficient, that the *Root* hath *Earth* to shoot into, or that its *Motion* is only an *Appetite* of being therein lodged, which way soever that be : but that its nature is, though within the *Earth* already, yet to change its *Position*, and to *move Downwards*. And so likewise of the *Trunk*, that it rises, when a *Seed* sprouts, out of the Ground, not meerly because it hath an *Appetite* of being in the open *Aer* ; for in this Experiment it is so already ; yet now makes a new *Motion* upwards."

Although Grew cannot be called a great experimenter, he frequently took the easier course of throwing out suggestions for such work. " The generation of Experiments " he describes as " being like that of Discourse, where one thing introduceth an hundred more which otherwise would never have been thought of." Amongst other proposals he recommends that trial should be made of growing plants in common water, snow

water, milk, oil, wine, ink, etc., or in any of these with solid
bodies, such as nitre and salt, dissolved in them. He points
out that the effect both on the plant and on the liquid should
be noted. The solid body should be weighed before solution,
and then, after the experiment is over, the liquid should be
evaporated and the solid again weighed.

Another instance in which he suggested an experiment,
apparently without carrying it out, was in relation to the movements of the stems of non-climbing plants. He seems to have
anticipated the nineteenth century discovery of nutation amongst
plants other than climbers, though he stopped short of actually
proving it. In his account of the Motions of Trunks he remarks,
" The *Convolution* of *Plants*, hath been observed only in those
that Climb. But it seems probable, that many others do
also *wind*;...Whether it be so, or not the Experiment may
easily be made by tying a *Thred* upon any of the *Branches* ;
setting down the respect it then hath to any Quarter in the
Heavens : for, if it shall appear in two or three Months, to have
changed its Situation towards some other Quarter ; it is certain
proof hereof." He noticed that some plants twine " by *South*
from *East* to *West*" and others " from *West* to *East*," and attributed this to their being respectively under the influence of the
sun and the moon.

Whenever Grew's notions of plant physiology depended
upon chemistry, they became, according to our modern ideas,
extremely difficult to follow. He held, among many other curious
beliefs, that salts obtained from any plant have a tendency to
crystallise out in a form resembling that plant, and adds, as
an illustration from the animal world, " though I have not seen
it my self, yet I have been told by one that doth not use to
phancy things, that the Volatile Salt of *Vipers*, will figure it self
into the semblance of little *Vipers*."

The mystical belief that characteristic " principles " permeate
all things, finds expression in his idea that the " frost flowers,"
sometimes to be seen on a window pane, are evidence that the
air is impregnated with " *Vegetable Principles*." Another fact,
which he brings forward in support of the same view, is that
the ground or water, when exposed for some time to air,

turns green. His explanation, in this latter case, was not far from the truth, for, as we now know, the greenness is due to the vegetation of minute algæ, which, in their dormant state, may be carried from place to place by the wind.

It is usual to regard Ecology as a very recent development of botanical science, but Nehemiah Grew seems to have been alive to the importance of the ecological standpoint,—though he did not describe it by this name. He writes "The proper *Places* also of *Plants*, or such wherein they have…a Spontaneous growth, should be considered. And that as to the *Climate*; whether in one Colder, Temperate, or more Hot. The *Region*; Continent, or Island. The *Seat*; as Sea, or Land, Watry, Boggy, or Dry; Hills, Plains, or Vallies; Open, in Woods, or under Hedges; against *Walls*, rooted in them, or on their Tops; and the like."

Grew's most interesting contribution to science was, perhaps, his publication of the fact that the flowering plants, like animals, shew the phenomena of sex. He never, however, actually proved this contention in an experimental way. At the time that his earliest work[1] was published, he was frankly puzzled by the stamens, or, as he calls them, the "Attire." He recognised their use to ·insects, to whom flowers serve, in his own words, as "their Lodging and their Dining-Room." He also fully realised their value to man as increasing the beauty of the blossom, but he was broad-minded enough to feel that these must be secondary uses, and that "the primary and private use of the *attire*" remained to be discovered. Ten years later, in the second edition of his work, he tells us that it was suggested to him in conversation by Sir Thomas Millington that the stamens were the male organs. It seems probable that, although Grew gives Millington the credit for this discovery, he had really arrived at it independently, for he tells us that when Millington made the suggestion, he "immediately reply'd that [he] was of the same Opinion; and gave him some reasons for it, and answered some *Objections*, which might oppose them."

Besides his belief in the male nature of the stamen, Nehemiah Grew came to some rather mysterious conclusions as to their

[1] *The Anatomy of Vegetables Begun*, 1672.

serving to draw off the redundant part of the sap, not needed to produce the seed. He also used the word "attire" for the florets of the Compositæ, but qualified it by calling the stamens the "seminal attire," and the florets of compound flowers the "florid attire." He says that "every *Flower* with the *Florid attire*" (or, as we should now say, "every composite flower") "Embosomes, or is, a *Posy* of perfect *Flowers*." He recognised the "globulets" (pollen grains) as being of the same nature as those in the anthers of simple flowers. He describes the disk florets with remarkable accuracy, but falls into the error of supposing that the pollen grains are in some cases originally produced by the style and stigmas, which he calls the "Blade," and which he did not recognise as part of the female organ. His figures make it clear that he mistook the stylar hairs for little stalks organically connecting the pollen grains and the style. In other cases, however, he observed that the pollen grains occurred on the inner side of what we now know as the staminal tube.

Grew enters into considerable detail as regards the structure of flowers, and it is only possible to mention here a few of the points to which he draws attention. He observed the frequent occurrence of capitate glandular hairs, which he describes as "like so many little *Mushrooms* sprouting out of·the *Flower*," their heads sometimes exuding a "*Gummy* or *Balsamick Juyce*." He describes the varieties of aestivation of the floral leaves, and notes that, in the Poppy, the large size and fewness of the petals prevents their being folded into a compact body by any of the ordinary methods. "For which reason, they are cramb'd up within the *Empalement*[1] by hundreds of little *Wrinckles* or *Puckers*; as if Three or Four Fine *Cambrick Handcherchifs* were thrust into ones *Pocket*."

We have said something about Grew's work on seeds, in dealing with his first treatise. He was always much interested in this subject, and returned to it again in his later work. He mentions the mucilaginous testa possessed by many seeds, but which only becomes noticeable when they have been moistened. That of "*Nasturtium Hortense*" he describes as very large, "even emulous of the inner Pulp surrounding a *Gooseberry-Seed*." He

[1] Calyx.

suggests that the value of putting a Clary seed into the eye
to bring out a foreign body, which may have lodged there, is
due to the presence of the mucilaginous coat. The same seed
is still, I believe, used for this purpose, under the name of "eye
seed." Grew understood the difference between seeds with, and
without endosperm, and gives perfectly clear representations of
such albuminous seeds as *Ricinus*. He describes the cotyledons
of the Dock as being immersed in the endosperm, "as in a *Tub*
of *Meal* or a little pot of pure refin'd *Mould*, necessary for the
first *Vegetation* of the *Radicle*."

Grew naturally reckoned the spores of Ferns among seeds.
The seed-case of the Harts-tongue is, he says, "of a *Silver
Colour*...of a *spherick Figure*, and girded about with a sturdy
Tendon or *Spring*, of the *Colour* of *Gold*:... So soon as...
this *Spring* is become stark enough, it suddenly breaks the
Case into two halfs, like two little *cups*, and so flings the *Seed*,"
of which "ten Thousand are not so big as a white *Pepper
Corn*."

To give any kind of short summary of Grew's botanical
work is well-nigh impossible. Some men are remembered for
individual discoveries, and in such cases it is not difficult to
give a précis of their contributions. But Nehemiah Grew is
remembered because, contemporaneously with Malpighi, he
actually created the science of plant anatomy,—a subject which,
before his day, was practically non-existent. Modern botanists,
conscious how small an addition to the fabric is now regarded
as a satisfactory life-work, must stand amazed and somewhat
humbled before the broad and sound foundations laid by this
seventeenth century physician. It is no less than two hundred
and forty years since Grew sent in his first treatise to the
Royal Society, so it is scarcely wonderful that a number of his
results have been rejected in course of time. It is far more
remarkable that so many of his conclusions—and those the
more essential ones—have been merely confirmed and extended
by later work. Great however as were his actual contributions
to botanical knowledge, they were perhaps less important than
the far-reaching service which he rendered in helping to free
biological thought from the cramping belief that the one and

only object of the existence of the organic world was for the use and pleasure of man. Grew believed that the "Outward Elegancies of *Plants*" might be for the purpose of giving delight to the human race, but he was the first to point out that as the "Inward Ones, which, generally, are as Precise and Various as the Outward," are so seldom seen, their purpose can hardly be for this, but must be for the benefit of the plants themselves, "That the *Corn* might grow, *so*; and the *Flower*, *so*, whether or no Men had a mind, leisure, or ability, to understand *how*."

Plate VIII

STEPHEN HALES (1759)

STEPHEN HALES

1677—1761

By FRANCIS DARWIN

An error corrected—Hales' scientific contemporaries—Physiology or Physics
—Hales the Founder of the experimental method in Physiology—His
style—Cambridge days—Teddington—*Vegetable Staticks*—Experiments
described — Transpiration — Root Pressure — Assimilation — Practical
application to greenhouses—Distribution of growth first measured—
Hales' other activities—Sachs' tribute.

IN attempting to give a picture of any man's life and work
it is well to follow the rule of the *Dictionary of National
Biography*, and begin with the dates of his birth and death.
Stephen Hales was born in 1677 and died in 1761, having had
experiences of the reigns of seven sovereigns.

The authorities for the life of Hales are given in my article
on him in the *Dictionary of National Biography*. Botanists in
general probably take their knowledge of the main facts of his
life from Sachs' *History of Botany*. It is therefore worth while
to point out that both the original and the English translation
(1890) contain the incorrect statement that Hales was educated
at Christ's College, Cambridge, and that he held the living of
Riddington, whereas he is one of the glories of Corpus, and
was perpetual curate of Teddington. These inaccuracies however are trifles in relation to the great and striking merits of
Sachs' *History*, a work which to my thinking exhibits the
strength and brilliance of the author's mind as clearly as any
of his more technical writings. Sachs was no niggling biographer, and his broad vigorous outlines must form the basis of
what anyone, who follows him, has to say about the Botanists
of a past day.

To return to Hales' birth: it is of interest to note how he fits into the changing procession of lives, to see what great men overlap his youth, who were his contemporaries in his maturity, and who were appearing on the scientific stage as he was leaving it.

Sir Isaac Newton was the dominant figure in English science while Hales was developing. He died in 1727, the year in which Hales published his *Vegetable Staticks*, a book, which like the *Origin of Species*, appeared when its author was 50 years of age; Newton was at the zenith of his fame when Hales was a little boy of 10—his *Principia* having been published in 1687. And when Hales went up to Cambridge in 1696 he must have seen the great man coming from his rooms[1] in the N.E. corner of the Great Court of Trinity—that corner where Newton's and other more modern ghosts surely walk—Macaulay who used to read, pacing to and fro by the chapel[2], and Thackeray who, like his own Esmond, lived "near to the famous Mr Newton's lodgings." In any case there can be no doubt that the genius of Newton cast its light on Hales, as Sachs has clearly pointed out (*Hist. Bot.*, Eng. Tr., p. 477). Another great man who influenced Hales was Robert Boyle, who was born 1627 and died 1691. John Mayow again, that brilliant son of Oxford, whose premature death at 39 in 1679 was so heavy a blow to science, belongs to the same school as Hales—the school which was within an ace of founding a rational chemistry, but which was separated from the more obvious founders of that science by the phlogiston-theory of Becchers and Stahl. I do not find any evidence that Hales was influenced by the phlogistic writers and this is comprehensible enough, if, as I think, he belongs to the school of Mayow and Boyle.

The later discoverers in chemistry are of the following dates, Black 1728—1799, Cavendish 1731—1810, Priestley 1733—1804, Scheele 1742—1786, Lavoisier 1743, guillotined 1794.

---

[1] In 1699 Newton was made master of the Mint and appointed Whiston his Deputy in the Lucasian Professorship, an office he finally resigned in 1703 (Brewster's *Life of Newton*, 1831, p. 249).

[2] "There, if anywhere, his dear shade must linger," Trevelyan, *Life and Letters of Lord Macaulay*, 1 volume edit. 1881, p. 55.

These were all born about the time of Hales' zenith, nor did he live[1] to see the great results they accomplished. But it should not be forgotten that Hales' chemical work made more easy the triumphant road they trod.

I have spoken of Hales in relation to chemists and physicists because, though essentially a physiologist, he seems to me to have been a chemist and physicist who turned his knowledge to the study of life, rather than a physiologist who had some chemical knowledge.

Whewell points out in his *History of the Inductive Sciences*[2] that the Physiologist asks questions of Nature in a sense differing from that of the Physicist. The *Why?* of the Physicist meant *Through what causes?* that of the Physiologist—To what end ? This distinction no longer holds good, and if it is to be applied to Hales it is a test which shows him to be a physicist. For, as Sachs shows, though Hales was necessarily a teleologist in the theological sense, he always asked for purely mechanical explanations. He was the most unvitalistic of physiologists, and I think his explanations suffered from this cause. For instance, he seems to have held that to compare the effect of heat on a growing root to the action of the same cause on a thermometer[3] was a quite satisfactory proceeding. And there are many other passages in *Vegetable Staticks* where one feels that his speculations are too heavy for his knowledge.

Something must be said of Hales' relation to his predecessors and successors in Botanical work. The most striking of his immediate predecessors were Malpighi 1628—1694, Grew 1628—1711, Ray 1627—1705, and Mariotte (birth unknown, died 1684) ; and of these the three first were born one hundred years before the publication of *Vegetable Staticks*. Malpighi and Grew were essentially plant-anatomists, though both dealt in physiological speculations. Their works were known to Hales, but they do not seem to have influenced him.

---

[1] Black's discovery of CO_2, however, was published in 1754, seven years before Hales died, but Priestley's, Cavendish's and Lavoisier's work on O and H was later.

[2] 1837, III. p. 389.

[3] *Vegetable Staticks*, p. 346.

We have seen that as a chemist Hales is somewhat of a solitary figure, standing between what may be called the periods of Boyle and of Cavendish. This is even more striking in his Botanical position, for here he stands in the solitude of all great original inquirers. We must go back to Van Helmont, 1577—1644, to find anyone comparable to him as an experimentalist. His successors have discovered much that was hidden from him, but consciously or unconsciously they have all learned from him the true method and spirit of physiological work.

It may be urged that in exalting Hales I am unfair to Malpighi. It may be fairer to follow Sachs in linking these great men together and to insist on the wonderful fact that before Malpighi's book in 1671, vegetable physiology was still where Aristotle left it, whereas 56 years later in 1727 we find in Hales' book an experimental science in the modern sense.

It should not be forgotten that students of animal physiology agree with botanists as to Hales' greatness. A writer in the *Encyclopædia Britannica* speaks of him as "the true founder of the modern experimental method in physiology."

According to Sachs, Ray made some interesting observations on the transmission of water, but on the whole what he says on this subject is not important. There is no evidence that he influenced Hales.

Mariotte the physicist came to one physiological conclusion of great weight[1]; namely, that the different qualities of plants, e.g. taste, odour, etc., do not depend on the absorption from the soil of differently scented or flavoured principles, as the Aristotelians imagined, but on *specific differences* in the way in which different plants deal with identical food material—an idea which is at the root of a sane physiological outlook. These views were published in 1679[2], and may have been known to Hales. He certainly was interested in such ideas, as is indicated by his attempts to give flavour to fruit by supplying them with medicated fluids. He probably did not expect success for he remarks, p. 360: "The specifick differences of vegetables, which are all sustained and grow from the same nourishment, is

[1] Sachs, *Geschichte*, p. 502.   Malpighi held similar views.
[2] *Ibid.*, p. 499.

[*sic*] doubtless owing to the very different formation of their minute vessels, whereby an almost infinite variety of combinations of the common principles of vegetables is made." He continues in the following delightful passage : "And could our eyes attain to a sight of the admirable texture of the parts on which the specific differences in plants depends [*sic*] what an amazing and beautiful scene of inimitable embroidery should we behold? what a variety of masterly strokes of machinery? what evident marks of consummate wisdom should we be entertained with?" To conclude what has been said on Hales' chronological position —Ingenhousz, the chief founder of the modern point of view on plant nutrition, was born 1730 and published his book *On Vegetables*, etc. in 1779. So that what was said of Hales' chemical position is again true of him considered in relation to nutrition ; he did not live to see the great discoveries made at the close of the 18th century.

There is in his writing a limped truthfulness and simplicity, unconsciously decorated with pretty 18th century words and half-rusticities which give it a perennial charm. And inasmuch as I desire to represent Hales not merely as a man to be respected but also to be loved, it will be as well to give what is known of the personal side of his character before going on to a detailed account of his work.

He was, as we have seen, entered at Corpus Christi College, Cambridge, in June, 1696. In February, 1702—3, he was admitted a fellow of the College. It was during his life as a fellow that he began to work at chemistry in what he calls "the elaboratory in 'Trinity College." The room is now occupied by the Senior Bursar and forms part of the beautiful range of buildings in the bowling green, which, freed from stucco and other desecration, are made visible in their ancient guise by the piety of a son of Trinity and the wisdom of the College authorities. It was here, according to Dr Bentley, that "the thieving Bursars of the old set embezzled the College timber[1]," and it was this room that was fitted up as "an elegant laboratory" in 1706 for John Francis

[1] Quoted by Caröe, in his paper read before the *Cambridge Archaeological Society* on *King's Hostel* etc., and "Printed for the Master and Fellows of Trinity Coll." in 1909.

Vigani, an Italian chemist, who had taught unofficially in the University for some years and became the first Professor of Chemistry at Cambridge in 1703.

Judging from his book, *Medulla Chymiae*, 1682, Vigani was an eminently practical person who cared greatly about the proper make of a furnace and the form of a retort, but was not cumbered with theories.

Hales vacated his fellowship and became minister or perpetual curate of Teddington[1] in 1708—9 and there he lived until his death, fifty-two years afterwards. He was married (? 1719) and his wife died without issue in 1721.

He attracted the attention of Royalty, and received plants from the King's garden at Hampton Court. Frederick Prince of Wales, the father of George III, is said to have been fond of surprising him in his laboratory at Teddington. This must surely be a unique habit in a prince, but we may remember that, in the words of the Prince's mock epitaph, "since it is only Fred there's no more to be said." He became Clerk of the Closet to the Dowager Princess and this " mother of the best of Kings" as she calls herself put up his monument in Westminster Abbey. Hales had the honour of receiving the Copley Medal from the Royal Society in 1739, and Oxford made him a D.D. in 1733.

Some years ago I made a pilgrimage to Teddington and found, in the parish registers, many interesting entries by his hand; the last in a tremulous writing is on November 4th, 1760, two months before he died. He was clearly an active parish priest. He made his female parishioners do public penance when he thought they deserved it: he did much for the fabric of the church. "In 1754[2] he helped the parish to a decent water supply and characteristically records, in the parish register, that the outflow was such as to fill a two-quart vessel in 'three swings of a pendulum beating seconds, which pendulum was $39 + \frac{2}{10}$ inches long from the suspending nail to the middle of the plumbet or bob'." Under the tower he helped

---

[1] He also held the living of Farringdon in Hampshire where he occasionally resided.

[2] *Dict. Nat. Biog.*

to build (which now serves as a porch) Stephen Hales is buried, and the stone which covers his body is being worn away by the feet of the faithful. By the piety of a few botanists a mural tablet, on which the epitaph is restored, has been placed near the grave. Horace Walpole called Hales "a poor, good, primitive creature" and Pope[1] (who was his neighbour) said "I shall be very glad to see Dr Hales, and always love to see him, he is so worthy and good a man." Peter Collinson writes of "his constant serenity and cheerfulness of mind"; it is also recorded that "he could look even upon wicked men, and those who did him unkind offices, without any emotion of particular indignation; not from want of discernment or sensibility; but he used to consider them only like those experiments which, upon trial, he found could never be applied to any useful purpose, and which he therefore calmly and dispassionately laid aside."

Hales' work may be divided into three heads:

    I.   Physiological, animal and vegetable;

    II. Chemical;

    III. Inventions and miscellaneous essays.

Under No. I. I shall deal only with his work on plants. The last heading (No. III.) I shall only refer to slightly, but the variety and ingenuity of his miscellaneous publications is perhaps worth mention here as an indication of the quality of his mind. It seems to me to have had something in common with the versatile ingenuity of Erasmus Darwin and of his grandson Francis Galton. The miscellaneous work also exhibits Hales as a philanthropist, who cared passionately for bettering the health and comfort of his fellow creatures by improving their conditions of life.

His chief book from the physiological and chemical point of view is his *Vegetable Staticks*. It will be convenient to begin with the physiological part of this book, and refer to the chemistry later. *Vegetable Staticks* is a small 8vo of 376 pages, dated on the title-page 1727. The "*Imprimatur* Isaac Newton Pr. Reg. Soc." is dated February 16, 172⅞, and this date is of

---

[1] With a certain idleness Pope reduces him to plain Parson Hale, for the sake of a rhyme in the *Epistle of Martha Blount*, 1. 198.

some slight interest, for Newton died on March 20, and *Vegetable Staticks* must have been one of the last books he signed.

The dedication is to George Prince of Wales, afterwards George III. The author cannot quite avoid the style of his day, for instance: "And as *Solomon* the greatest and wisest of men, deigned[1] to inquire into the nature of Plants, *from the Cedar of Lebanon, to the Hyssop that springeth out of the wall.* So it will not, I presume, be an unacceptable entertainment to your Royal Highness," etc.

But the real interest of the dedication is its clear statement of his views on the nutrition of plants. He asserts that plants obtain nourishment, not only from the earth, "but also more sublimed and exalted food from the air, that wonderful fluid, which is of such importance to the life of Vegetables and Animals," etc. We shall see that his later statement is not so definite, and it is well to rescue this downright assertion from oblivion.

His book begins with the research for which he is best known, namely that on transpiration. He took a sunflower growing in a flower-pot, covering the surface of the earth with a plate of thin milled lead, and cemented it so that no vapour could pass, leaving a corked hole to allow of the plant being watered. He did not take steps to prevent loss through the pot, but at the end of the experiment cut off the plant, cemented the stump and found that the "unglazed porous pot" perspired 2 ozs. in 12 hours, and for this he made due allowance.

The plant so prepared he proceeded to weigh at stated intervals. He obtained the area of the leaves by dividing them into parcels according to their several sizes and measuring one leaf[2] of each parcel. The loss of water in 12 hours converted to the metric system is 1·3 c.c. per 100 sq. cm. of leaf-surface ; and this is of the same order of magnitude as Sachs' result[3], namely 2·2 c.c. per 100 sq. cm.

[1] The original reads "deigned not," an obvious slip.
[2] This he does by means of a network of threads ¼ inch apart. Pfeffer, *Pflanzenphysiologie*, ed. 1, 1. p. 142, recommends the method and gives Hales as his authority.
[3] *Pflanzenphysiologie*, 1865 (Fr. Trans. 1868), p. 254.

He goes on to measure the surface of the roots[1] and to estimate the rate of absorption per area. The calculation is of no value, since he did not know how small a part of the roots is absorbent, nor how enormously the surface of that part is increased by the presence of root-hairs. He goes on to estimate the rate of the flow of water up the stem; this would be 34 cubic inches in 12 hours if the stem (which was one square inch in section) were a hollow tube. He then allowed a sunflower stem to wither and to become completely dry, and found that it had lost ¾ of its weight, and assuming that the ¼ of the "solid parts" left was useless for the transmission of water he increases his 34 by ⅓ and gives 45⅓ cubic inches in 12 hours as the rate. But the solid matter which he neglected contained the vessels and he would have been nearer to the truth had he corrected his figures on this basis. The simplest plan is to compare his results with those obtained by Sachs[2] in allowing plants to absorb solutions of lithium-salts. If the flow takes place through conduits equivalent to a quarter of a square inch in area, the fluid will rise in 12 hours to a height of 4 × 34 or 136 inches or in one hour to 28·3 cm.[3] This is a result comparable to, though very much smaller than, Sachs' result with the sunflower, viz. 63 cm. per hour.

The data are however hardly worth treating in this manner. But it is of historic interest to note that when Sachs was at work on his *Pflanzenphysiologie*, published in 1865, he was compelled to go back nearly 140 years to find any results with which he could compare his own.

We need not follow Hales into his comparison between the "perspiration" of the sunflower and that of a man, nor into his other transpiration experiments on the cabbage, vine, apple, etc. But one or two points must be noted. He found[4] the "middle rate of perspiration" of a sunflower in 12 hours of daylight to be 20 ounces, and that of a "dry warm night" about 3 ounces;

[1] He gives it as 15·8 square inches, the only instance I have come across of his use of decimals.
[2] *Arbeiten*, II. p. 182.
[3] See Sachs' *Pflanzenphys.* 1865 (Fr. Trans. 1868), p. 257, where the above correction is applied to Hales' work.
[4] *Vegetable Staticks*, p. 5.

thus the day transpiration was roughly seven times the nocturnal rate. This difference may be accounted for by the closure of the stomata at night.

Hales of course knew nothing of stomata, but it is surprising to find Sachs in 1865 discussing the problem of transpiration with hardly a reference to the effect of stomatal closure.

Hales[1] notes another point which a knowledge of stomatal behaviour might have explained, viz. that with "scanty watering the perspiration much abated," he does not attempt an explanation but merely refers to it as a "healthy latitude of perspiration in this Sunflower."

In the course of his work on sunflowers he notices that the flower follows the sun, he says however that it is "not by turning round with the sun," i.e. that it is not a twisting of the stalk, and goes on to call it *nutation* which must be the *locus classicus* for the term used in this sense.

An experiment[2] that I do not remember to have seen quoted elsewhere is worth describing. It is one of the many experiments that show the generous scale on which his work was planned. An apple bough five feet long was fixed to a vertical glass tube nine feet long. The tube being above and the branch hanging below the pressure of the column of water would act in concert with the suck of the transpiring leaves instead of in opposition to this force. He then cut the bare stem of his branch in two, placing the apical half of the specimen (bearing side branches and leaves) with its cut end in a glass vessel of water, the basal and leafless half of the branch remained attached to the vertical tube of water. In the next 30 hours only 6 ounces dripped through the leafless branch, whereas the leafy branch absorbed 18 ounces. This, as he says, shows the great power of perspiration. And though he does not pursue the experiment, it is worthy of note as an attempt like those of Janse[3] and others to correlate the flow of water under pressure with the flow due to transpiration.

It is interesting to find that Hales used the three methods of

[1] *Vegetable Staticks*, p. 14.   [2] *Vegetable Staticks*, p. 41.
[3] Janse in *Pringsheinis Jahrb.* XVIII. p. 38. The later literature is given by Dixon in *Progressus Rei Bot.* III., 1909, p. 58.

estimating transpiration which have been employed in modern times, namely, (i) weighing, (ii) a rough sort of potometer, (iii) enclosing a branch in a glass balloon and collecting the precipitated moisture, the well-known plan followed by various French observers.

He (*Vegetable Staticks*, p. 51) concluded his balance of loss and gain in transpiring plants by estimating the amount of available water in the soil to a depth of three feet, and calculating how long his sunflower would exist without watering. He further concludes (p. 57) that an annual rainfall (of 22 inches) is "sufficient for all the purposes of nature, in such flat countries as this about Teddington."

He constantly notes small points of interest, e.g. (p. 82) that with cut branches the water absorbed diminishes each day and that the former vigour of absorption may be partly renewed by cutting a fresh surface[1].

He also showed (p. 89) that the transpiration current can flow perfectly well from apex to base when the apical end is immersed in water.

These are familiar facts to us, but we should realise that it is to the industry and ingenuity of Hales that we owe them. In a repetition (p. 90) of the last experiment, we have the first mention of a fact fundamentally important. He took two branches (which with a clerical touch he calls *M* and *N*) and having removed the bark from a part of the branch dipped the ends in water, *N* with the great end downwards, but *M* upside down. In this way he showed that the bark was not necessary for the absorption or transmission of water[2]. I suspect that one branch was inverted out of respect for the hypothesis of sap-circulation. He perhaps thought that water could travel apically by the wood, but only by the bark in the opposite direction.

Later in his book (pp. 128 and 131) he gives definite arguments against the hypothesis in question.

Next in order (p. 95) comes his well-known experiment on the pressure exerted by peas increasing in size as they imbibe

---

[1] Compare F. von Höhnel, *Bot. Zeitung*, 1879, p. 318.

[2] This is also shown by experiment xc, *Vegetable Staticks*, p. 123.

water. There are, however, pitfalls in this result of which Hales was unaware, and perhaps the chief interest to us now is that he considered the imbibition of the peas[1] to be the same order of phenomenon as the absorption of water by a cut branch—notwithstanding the fact that he knew[2] the absorption to depend largely on the leaves. It may be noticed that Sachs with his imbibitional view of water-transport may be counted a follower of Hales.

In order to ascertain "whether there was any lateral communication of the sap and sap vessels, as there is of blood in animals," Hales (p. 121) made the experiment which has been repeated in modern laboratories[3], i.e. cutting a "gap to the pith" and another opposite to it and a few inches above. This he did on an oak branch six feet long whose basal end was placed in water. The branch continued to "perspire" for two days, but gave off only about half the amount of water transpired by a normal branch[4]. He does not trouble himself about this difference, being satisfied of "great quantities of liquor having passed laterally by the gap."

He is interested in the fact of lateral transmission in connexion with the experiment of the suspended tree (Fig. 24, p. 126), which is dependent on the neighbours to which it is grafted for its water supply. This seems to be one of the results that convinced him that there is a distribution of food material which cannot be described as circulation of sap in the sense that was then in vogue.

Hales (p. 143) was one of the first[5] to make the well-known experiment—the removal of a ring of bark, with the result that the edge of bark nearest the base of the branch swells and thickens in a characteristic manner. He points out that if a number of rings are made one above the other, the swelling is seen at the lower edge of each isolated piece of bark, and

---

[1] The method by which Hales proposed to record the depth of the sea is a variant of this apparatus.

[2] *Vegetable Staticks*, p. 92.

[3] According to Sachs (*Geschichte*, p. 509) Ray employed this method.

[4] Other facts show that the "gapped" branches did not behave quite normally.

[5] He refers (p. 141) to what is in principle the same experiment (see Fig. 27) as due to Mr Brotherton, and published in the *Abridgment of the Phil. Trans.* II. p. 708.

therefore (p. 143) the swelling must be attributed "to some other cause than the stoppage of the sap in its return downwards," because the first gap in the bark should be sufficient to check the whole of the flowing sap[1]. He must in fact have seen that there is a redistribution of plastic material in each section of bark.

We now for the moment leave the subject of transpiration and pass on to that of root-pressure on which Hales is equally illuminating.

Figure from *Vegetable Staticks* showing a vine with mercury gauges in place to demonstrate root-pressure.

His first experiment, *Vegetable Staticks*, p. 100, was with a vine to which he attached a vertical pipe made of three lengths of glass-tubing jointed together. His method is worth notice. He attached the stump to the manometer with a " stiff cement made of melted Beeswax and Turpentine, and bound it over with several folds of wet bladder and pack-thread." We cannot wonder that the making of water-tight connexions was a great difficulty, and we can sympathise with his belief that he could have got a column more than 21 feet high but for the leaking of the joints on several occasions. He notes the

---

[1] He notices that the swelling of the bark is connected with the presence of buds. The only ring of bark which had no bud showed no swelling.

familiar fact that the vine-stump absorbed water before it began to extrude it.

He afterwards (pp. 106–7) used a mercury gauge and registered a root-pressure of 32¼ inches or 36 feet 5⅛ inches of water which he proceeds to compare with his own determination of the blood-pressure of the horse (8 feet) and of other animals. Perhaps the most interesting of his root-pressure experiments was that (p. 110) in which several manometers were attached to the branches of a bleeding vine and showed a result which convinced him that "the force is not from the root only, but must proceed from some power in the stem and branches," a conclusion which some modern workers have also arrived at. The figure on page 77 is a simplified reproduction of the plate (Fig. 19) in *Vegetable Staticks*.

Assimilation.

Hales' belief that plants draw part of their food from the air, and again that air is the breath of life, of vegetables as well as of animals (p. 148), are based upon a series of chemical experiments performed by himself. Not being satisfied with what he knew of the relation between "air" (by which he meant gas) and the solid bodies in which he supposed gases to be fixed, he delayed the publication of *Vegetable Staticks* for some two years, and carried out the series of observations which are mentioned in his title-page as " An attempt to analyse the air, by a great variety of chymio-statical experiments" occupying 162 pages of his book[1].

The theme of his inquiry he takes (*Vegetable Staticks*, p. 165) from " the illustrious Sir *Isaac Newton*," who believed that "Dense bodies by fermentation rarify into several sorts of Air ; and this Air by fermentation, and sometimes without it, returns into dense bodies."

Hales' method consisted in heating a variety of substances, e.g. wheat-grains, pease, wood, hog's blood, fallow-deer's horn, oyster-shells, red-lead, gold, etc., and measuring the "air" given off from them. He also tried the effect of acid on iron filings,

[1] It appears that Mayow made similar experiments. *Dict. Nat. Biog.* s.v. Mayow.

oyster-shells, etc. In the true spirit of experiment he began by strongly heating his retorts (one of which was a musket barrel) to make sure that no air arose from them. It is not evident to me why he continued at this subject so long. He had no means of distinguishing one gas from another, and almost the only quality noted is a want of permanence, e.g. when the CO_2 produced was dissolved by the water over which he collected it. Sir E. Thorpe[1] points out that Hales must have prepared hydrogen, carbonic acid, carbonic oxide, sulphur dioxide, marsh gas, etc. It may, I think, be said that Hales deserved the title usually given to Priestley, viz. " the father of pneumatic[2] chemistry."

Perhaps the 'most interesting experiment made by Hales is the heating of minium (red-lead) with the production of oxygen. It proves that he knew, as Boyle, Hooke and Mayow did before him, that a body gains weight in oxidation. Thus Hales remarks: "That the sulphurous and aereal particles of the fire are lodged in many of those bodies which it acts upon, and thereby considerably augments their weight, is very evident in Minium or Red Lead which is observed to increase in weight in undergoing the action of the fire. The acquired redness of the Minium indicating the addition of plenty of sulphur in the operation." He also speaks of the gas distilled from minium, and remarks " It was doubtless this quantity of air in the minium which burst the hermetically sealed glasses of the excellent Mr *Boyle*, when he heated the Minium contained in them by a burning glass " (p. 287).

This was the method also used by Priestley in his celebrated experiment of heating red-lead in hydrogen ; whereby the metallic lead reappears and the hydrogen disappears by combining with the oxygen set free. This was expressed in the language of the day as the reconstruction of metallic lead by the addition of phlogiston (the hydrogen) to the calx of lead (minium). Thorpe points out the magnitude of the discovery that Priestley missed, and it may be said that Hales too was on

[1] *History of Chemistry*, 1909, I. p. 69.
[2] Hales made use of a rough pneumatic trough, the invention of which is usually ascribed to Priestley (Thorpe's *History of Chemistry*, I. p. 79).

the track and had he known as much as Priestley it would not have been phlogiston that kept him from becoming a Cavendish or Lavoisier. What chiefly concerns us however is the bearing of Hales' chemical work on his theories of nutrition. He concludes that "air makes a very considerable part of the substance of Vegetables," and goes on to say (p. 211) that "many of these particles of air" are "in a fixt state strongly adhering to and wrought into the substance of" plants[1]. He has some idea of the instability of complex substances and of the importance of the fact, for he says[2] that "if all the parts of matter were only endued with a strongly attracting power, [the] whole [of] nature would then become one unactive cohering lump." This may remind us of Herbert Spencer's words: "Thus the essential characteristic of living organic matter, is that it unites this large quantity of contained motion with a degree of cohesion that permits temporary fixity of arrangement," *First Principles*, § 103. With regard to the way in which plants absorb and fix the "air" which he finds in their tissues, Hales is not clear; he does not in any way distinguish between respiration and assimilation. But as I have already said he definitely asserts that plants draw "sublimed and exalted food" from the air.

As regards the action of light on plants, he suggests (p. 327) that "by freely entering the expanded surfaces of leaves and flowers" light may "contribute much to the ennobling principles of vegetation." He goes on to quote Newton (*Opticks, query* 30): "The change of bodies into light, and of light into bodies is very conformable to the course of nature, which seems delighted with transformations." It is a problem for the antiquary to determine whether or no Swift took from Newton the idea of bottling and recapturing sunshine as practised by the philosopher of Lagado. He could hardly have got it from Hales since *Gulliver's Travels* was published in 1726, a year before *Vegetable Staticks*. Timiriazeff, in his Croonian Lecture[3], was the first to see the connexion between photosynthesis and the Lagado research.

---

[1] He speaks here merely of the apples used in a certain experiment, but it is clear that he applies the conclusion to other plants.

[2] *Vegetable Staticks*, p. 313. It should be noted that Hales speaks of organic as well as inorganic substances.

[3] *Proc. R. Soc.* LXXII., p. 30, 1903.

Nevertheless Hales is not quite consistent about the action of light; thus (p. 351) he speaks of the dull light in a closely planted wood as checking the perspiration of the lower branches so that "drawing little nourishment, they perish." This is doubtless one effect of bad illumination under the above-named conditions, but the check to photosynthesis is a more serious result. In his final remarks on vegetation (p. 375) Hales says in relation to greenhouses, "it is certainly of as great importance to the life of the plants to discharge that infected rancid air by the admission of fresh, as it is to defend them from the extream cold of the outward air." This idea of ventilating greenhouses he carried out in a plant house designed by him for the Dowager Princess of Wales, in which warm fresh air was admitted. The house in question was built in 1761 in the Princess's garden at Kew, which afterwards became what we now know as Kew Gardens. The site of Hales' greenhouse, which was only pulled down in 1861, is marked by a big *Wistaria* which formerly grew on the greenhouse wall. It should be recorded that Sir W. Thiselton-Dyer[1] planned a similar arrangement independently of Hales, and found it produced a marked improvement of the well-being of the plants.

It is an illuminating fact that though Hales must have known Malpighi's theory of the function of leaves (which was broadly speaking the same as his own), he does not as far as I know refer to it. In his preface, p. ii, he regrets that Malpighi and Grew, whose anatomical knowledge he appreciated, had not "fortuned to have fallen into this *statical*[2] way of inquiry." I believe he means an inquiry of an experimental nature, and I think it was because Malpighi's theory was dependent on analogy rather than on ascertained facts, that it influenced Hales so little.

There is another part of physiology on which Hales threw light. He was the first I believe to investigate the distribution

[1] The above account of Hales' connexion with the Royal Gardens at Kew is from the *Kew Bulletin*, 1891, p. 289.

[2] I am indebted to Sir E. Thorpe for a definition of *statical*. "*Statical* (Med.) noting the physical phenomena presented by organised bodies in contradiction to the organic or vital." (Worcester's *Dictionary*, 1889.)

of growth in developing shoots and growing leaves by marking
them and measuring the distance between the marks after an
interval of time. He describes (p. 330) and figures (p.
344) with his usual thoroughness the apparatus employed: this was
a comb-like object, shown in Plate IX, made by fixing five pins
into a handle, ¼ inch apart from one another: the points being
dipped in red-lead and oil, a young vine-shoot was marked with
ten dots ¼ inch apart. In the autumn he examined his specimen
and finds that the youngest internode or "joynt" had grown
most, and the basal part having been "almost hardened" when
he marked, had "extended very little." In this—a tentative experiment—he made the mistake of not re-measuring his plants
at short intervals of time, but it was an admirable beginning and
the direct ancestor of Sachs'[1] great research on the subject.

In his discussion on growth it is interesting to find the idea
of turgescence supplying the motive force for extension. This
conception he takes from Borelli[2].

Hales sees in the nodes of plants "plinths or abutments for
the dilating pith to exert its force on" (p. 335); but he acutely
foresees a modern objection[3] to the explanation of growth as
regulated solely by the hydrostatic pressure in the cell. Hales
says (p. 335): "but a dilating spongy substance, by equally
expanding itself every way, would not produce an oblong shoot,
but rather a globose one."

It is not my place to speak of Hales' work in animal physiology, nor of those researches bearing on the welfare of the human
race which occupied his later years. Thus he wrote against the
habit of drinking spirits, and made experiments on ventilation
by which he benefited both English and French prisons, and even
the House of Commons; then too he was occupied in attempts
to improve the method of distilling potable water at sea, and of
preserving meat and biscuit on long voyages[4].

We are concerned with him simply as a vegetable physiologist

[1] *Arbeiten*, I.
[2] Borelli, *De Motu Animalium*, Pt II. Ch. xiii. According to Sachs, *Ges. d. Botanik*, p. 582, Mariotte (1679) had suggested the same idea.
[3] Nägeli, *Stärkekorner*, p. 279.
[4] See his *Philosophical Experiments*, 1739.

Plate IX

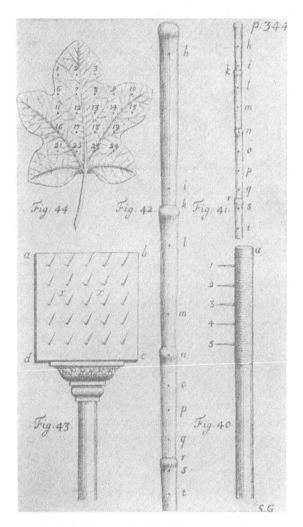

Plate 18 from Hales's *Vegetable Staticks*

Fig. 40. Instrument devised by Hales to make prick-
marks on a young shoot of Vine (Fig. 41); the
distribution of stretching after growth is shown in
Fig. 42. The use of a similar instrument for
marking surfaces is shown in Figs. 43 and 44

and in that character his fame is imperishable. Of the book which I have been using as my text, namely, *Vegetable Staticks*, Sachs says: "It was the first comprehensive work the world had seen which was devoted to the nutrition of plants and the movement of their sap....Hales had the art of making plants reveal themselves. By experiments carefully planned and cunningly carried out he forced them to betray the energies hidden in their apparently inactive bodies[1]." These words, spoken by a great physiologist of our day, form a fitting tribute to one who is justly described as the father of physiology.

[1] *Geschichte d. Botanik*, p. 515 (free translation).

JOHN HILL

1716—1775

BY T. G. HILL

Narrative—chequered career—journalism—attack on the Royal Society—
literary activities—Botanical works—structure of Timber—the sleep
of Plants—Mimosa and Abrus—views on Pollen—Hill's *Herbal*—his
admiration of Linnaeus—with qualifications—Hill's *Vegetable System*—
an ambitious work—financial losses—estimate of Hill's character.

IT has recently been remarked that the number of the
biographies of eminent men is inversely proportional to the
known facts concerning them. Although this generalisation is
probably incorrect, it is, to a certain extent, true of John Hill;
for, although he finds a place in biographical dictionaries,
apparently no extended account of his life has appeared. This
is a little surprising since, apart from his scientific work, he
occupied a prominent position in the middle of the eighteenth
century.

John Hill was the second son of the Rev. Theophilous Hill,
and was born either at Spalding or at Peterborough in the year
1716 or 1717. Nothing appears to be known regarding his
early education; according to Hawkins[1] he did not receive an
academical education, but there is no doubt that, as was usual
for those who desired to practise medicine at that and at much
later times, he served his apprenticeship to an apothecary, it is
said, at Westminster; also he attended the lectures on Botany
given under the auspices of the Apothecaries' Company at the
Chelsea Physic Garden. He first practised in St Martin's Lane

---

[1] Sir John Hawkins, *Life of Samuel Johnson*, London, 1787.

Plate X

JOHN HILL

in a shop which, according to Woodward[1], was little more than a shed; from there he moved to Westminster, and it appears that at the age of twenty-one he had a practice in Covent Garden. He early experienced financial difficulties; indeed, it is stated that, at times, he was unable to provide himself with the bare necessities of life. His marriage with a dowerless maiden, Miss Travers, did not improve his prospects, and he sought to add to his income by the utilization of his botanical knowledge. He travelled over the country collecting plants, which he dried, put up into sets with descriptions and sold by subscription; also he arranged the collections and gardens of the Duke of Richmond and Lord Petrie. Hill soon found that Botany, from the monetary point of view, was unprofitable; he therefore decided to try his fortune on the stage, and appeared at the Haymarket and Covent Garden.

Woodward[2] gives a very amusing account of him in his new profession. After giving examples to shew Hill's limitations, he remarks: " There was a time at the celebrated Theatre of *May Fair* he [Marr] represented *Altamont*, and the *Great Inspector* [Hill] attempted *Lothario*; and the polite Audience of that Place all choruss'd and agreed with you, when you dying, said, ' O Altamont! thy Genius is the stronger.'...Can I forget, great Sir, your acting *Constant*, in the *Provok'd Wife*, and your *innocent* Rape of Mrs Woffington; when, in a certain Passage, where, at least, a seeming Manliness was necessary, you handled her so awkwardly, that she joined the Audience in laughing at you."

Woodward's account may be accepted as being substantially correct, for in many ways Hill shewed that he lacked the qualities requisite for a successful career on the stage in those days.

Having thus failed as an actor, Hill returned to the practice of medicine and seemingly with more success, for in 1746 he was serving as a regimental surgeon, a position doubtless not very remunerative but helping to keep the wolf from the door. This same year saw the publication of Theophrastus's treatise on gems. In its new guise the value of the work was much

---

[1] A Letter from Henry Woodward,...to Dr Hill...London, 1752.
[2] *Loc. cit.*

enhanced since Hill intercalated much information that was lacking in the original ; further, the work was so well executed that it gained him the attention and good-will of eminent Fellows of the Royal Society.

The publication of this work was probably the turning point in Hill's career, and its success must have influenced him not a little in the determination of following a literary career. In 1846 he edited the *British Magazine*, a periodical which lived but four years. His activities in this direction were phenomenal, and it is hard to realize how he managed to find time for so much work, for in addition to his botanical publications, which will be considered hereafter, he wrote on such diverse subjects as the art of acting, the conduct of married life, theology, naval history, astronomy, entomology, human anatomy and other medical subjects. Also he wrote an opera, two farces, and certain novels. Much of this output represents mere hack work, but it shews that Hill had an enormous capacity for work, indeed on one occasion when he was sick, he confessed to a friend that he had overtaxed his strength in writing seven works at the same time.

The *Dictionary of National Biography* gives 76 titles of his publications, exclusive of eight which are generally attributed to him. Hill's output was probably even more extensive, for towards the latter part of his career he sometimes used to publish under a pseudonym. It is the more remarkable since he found time to enjoy the good things of the world, without which indulgence, according to his biographer[1], "he could not have undergone the fatigue and study inseparable from the execution of his vast designs." Again, according to Fitzgerald[2], he was "invariably in the front row at the theatres, exciting attention by his splendid dress and singular behaviour. When there was loud applause for the King, the doctor was seen to rise, and bow gravely to his Majesty."

The next few years were eventful ones for Hill. In 1751 he contributed a daily letter, called the *Inspector*, to the *London*

---

[1] *Short Account of the Life, Writings and Character of the late Sir John Hill, M.D.*, Edinburgh, 1779.

[2] Fitzgerald, *Life of Garrick*, London, 1868.

Advertiser and Literary Gazette; although they came to an end in 1753, the Inspectors were highly remunerative, thus it is stated that in one year Hill profited to the extent of £1500 by their sale, a very large sum for journalistic work in those days. They thus brought him very prominently before the public, and incidentally proved a source of some trouble to him.

In connexion with the *Inspector* justice has not been altogether done to Hill: no doubt, as Isaac Disraeli[1] states, that in them he retailed all the great matters relating to himself and all the little matters relating to others, but they were not all concerned in retailing the tales of scandal heard in the Coffee Houses and other places of public resort; nor were they always rendered palatable by these means as is stated in Rose's *Biographical Dictionary*[2]. They, in addition to comments and criticisms on current affairs, treated of many subjects. For instance, one considers the proposal for uniting the kingdoms of Great Britain and Ireland, another is a very sympathetic and laudatory review of Gray's *Elegy*, whilst a third treats of the art of embalming. Many are concerned with Natural History, and these are important as they shew Hill in another and very important character, namely that of a popular writer on Natural History, especially Botany. In one number he described the structure of a common flower, including an account of the movements of a bee in collecting pollen; and in another he described the appearance of microscopic organisms paying marked attention to their activities. These particular Inspectors are very pleasing and are well and clearly written; one especially is of outstanding importance, as it shews that Hill was in some respects far in advance of his times. He put forward a suggestion that Botany would be much improved by the delivery of public lectures in the museum with the living plants before the lecturer and the members of the audience. This scheme has yet to be carried out; as they are, museums are a means of education for the few, but a source of confusion to the many. For the latter their educative value would be enormously increased by the delivery

[1] Isaac Disraeli, *The Calamities and Quarrels of Authors*, London, 1865.
[2] London, 1848.

of lectures illustrated by the exhibits, for the spoken word is more abiding than the printed label.

The methods of criticism pursued by Hill in the *Inspector* soon involved him in controversy with various people. It is a difficult matter to appraise him in these respects ; possibly his success had turned his head for, according to Baker[1], he shewed " an unbounded store of vanity and self-sufficiency, which had for years lain dormant behind the mask of their direct opposite qualities of humility and diffidence ; a pride which was perpetually laying claim to homage by no means his due, and a vindictiveness which never could forgive the refusal of it to him." Baker then goes on to remark that as a consequence of this, every affront however slight was revenged by Hill by a public attack on the morals etc. of the maker.

On the other hand his criticisms may have been honest, at any rate in part ; and the fact that they landed him into difficulties does not necessarily indicate that he was a dishonest fellow ; most people are impatient of adverse criticism, and in those days such impatience found a vent in a pamphlet war or in personal violence. Nowadays the aggrieved manager, for instance, can shut his theatre doors against the distasteful critic ; or, in other cases, an action for libel appears to be not altogether unfashionable.

His attack on the Royal Society.

The real origin of Hill's attack on this learned society is somewhat obscure.

At the time of his death Chambers was engaged in the preparation of a supplement to his *Cyclopaedia*. The publishers then commissioned John Lewis Scott to prepare the work, but as Scott was soon afterwards appointed tutor to the royal princes it was entrusted to Hill. It is stated that the botanical articles were quite good, but that the more general parts were done with Hill's "characteristic carelessness and self-sufficiency." When the work was approaching completion the publishers considered that the title-page would look better if Hill had the right of

---

[1] *Biographica Dramatica,* 1812.

adding F.R.S. after his name. He, in consequence, and, it is stated, contrary to the advice of Folkes, endeavoured to obtain the necessary qualification for candidature; but he was disliked to such an extent that he could not obtain the requisite number of signatures, three, for his certificate, notwithstanding the fact that the number of Fellows was about three hundred. This perhaps was hardly surprising since he had criticized his contemporary scientists very adversely, designating them by such terms as "butterfly hunters," "cockle shell merchants" and "medal scrapers." This reverse must have been a severe blow to his vanity, for there can be no doubt that his claims to the Fellowship, on scientific grounds, were as strong as any and stronger than those of most of the Fellows. And this Hill, who was by no means lacking in self-confidence, knew. His criticism of the Society culminated in his *Review of the Works of the Royal Society of London* (1751)[1], which was in appearance like that of the *Transactions*, and consisted of reviews of several papers with comments by Hill. The work was dedicated to Martin Folkes, the President, on whom he placed the responsibility for publication, for, wrote he in his dedication, "The Purport of the more considerable of them has been long since delivered to you in conversation; and if you had thought the Society deserved to escape the Censure that must attend this Method of laying them before the World, you might have prevented it, by making the necessary Use of them in private.

"Nor is this, Sir, the only Sense in which you have been the great Instrument of their Production; since it cannot but be acknowledged, that if any body, except your great Self, had been in the high Office you so worthily fill at present, the Occasions of many of the more remarkable of them could not have been received by the Body, under whose Countenance alone they claim their Places in this Work."

He then charges Folkes with unworthy conduct towards him, and, in brief, he considered that Folkes and Baker were his enemies. The reason for this, according to Hill, was as follows.

---

[1] See also *Lucine sine concubitu*. A Letter addressed to the Royal Society, London, 1750. *A Dissertation on Royal Societies*, London, 1750.

An eminent French correspondent had taxed him, supposing him to be a Fellow, with "one of the errors of the Society"; Hill in reply wrote, "I have already set right the error you complain of; but you are to know, that I have the Honour not to be a Member of the Royal Society of London." Before he had sealed this letter he was called out of the room, and before he had returned a visitor, a Fellow of the Society, was shewn into Hill's study and read the letter containing the above-quoted passage. Hence the friction. Hill denies that he ever became a candidate for election, and states that although he attended the meetings he would not become a member on account of the Society's method of performing that which they were founded to do.

These statements are not lacking in definition; with regard to the incident of the letter it is impossible to judge of the truth; but with regard to the main features of the controversy the present writer thinks it extremely probable that the account first given is substantially correct, notwithstanding the statement that Hill's explanation was never contradicted[1].

As regards the *Review*, Hill wrote that "he pretends to nothing but the knowing more than the Royal Society of London appears by its publications to know! and surely a Man may do that and yet be very ignorant!"

The intention of the *Review* was to point out to the Society its shortcomings, doubtless in order that it might reform itself.

There can be no doubt whatever that a candid critic was necessary, for some of the papers were absolute rubbish, so much so indeed that a scientific training does not appear necessary to detect their futility. To take a brief example; in one paper the author describes a method to make trees grow very large; the seeds are to be sown at the absolute moment of the entry of the sun into the vernal equinox, and then to transplant them at the moment when the moon is full.

Hill himself sometimes falls into error in his criticisms; thus

---

[1] *Short Account of the Life, Writings and Character of the late Sir John Hill, M.D.*, Edinburgh 1779.

he adversely comments on the truth of the power of cobwebs to catch thrushes[1].

At the beginning of Part VII of his *Review*, which treats of plants, he thrusts very deep. He says, "This is a Branch of Natural Knowledge, which, it will appear, that the Royal Society of London have looked so very deeply into, that their rejecting the Linnean System of Botany, when offered by its Author will no longer be wondered at."

In this Part he is particularly severe upon Baker, and, in reading it, one is forced to the conclusion that although adverse criticism was warranted, there was a good deal of personal feeling behind it.

This attack on the Royal Society appears to have been much resented, and Hill's credit consequently was much damaged, for it was considered that Folkes and Baker had befriended him in his earlier days. With regard to Folkes it has been seen that Hill considered that he was doing a public duty; and with regard to Baker, Hill suffered under a real or imaginary grievance which, assuming Baker had helped him in the past, cancelled all obligations due from him to Baker. If this be not so then Hill, in addition to his other faults, was lacking in gratitude. With regard to this point his anonymous biographer[2] wrote that "we have nowhere learnt that ingratitude had the smallest share in the composition of the character of Sir John Hill."

The attack, however, was not altogether fruitless, as Disraeli[3] remarks, "Yet Sir John Hill, this despised man, after all the fertile absurdities of his literary life, performed more for the improvement of the Philosophical Transactions, and was the cause of diffusing a more general taste for the science of botany, than any other contemporary."

It is hardly necessary to remark that Hill was never elected to the Royal Society.

Thus by his methods of criticism Hill brought to an end a period of highly remunerative literary work; it was therefore necessary for him to seek other pastures. He returned, in part,

---

[1] See Bates, *The Naturalist on the River Amazons*. Edited by Ed. Clodd; London, 1892, p. 80. [2] *Short Life, loc. cit.* [3] *Loc. cit.*

to the practice of medicine in the shape of herbalist, preparing remedies from various plants such as valerian, water-dock and centaury; also he wrote on the virtues of these and other plants. The source from which he obtained his plants was in the first instance the Chelsea Physic Garden, but it is stated that he was eventually forbidden its use owing to his depredations; later he grew the requisite plants in his own garden which was situated where now is Lancaster Gate. There was a good deal of common sense in his remedies; thus in his *Virtues of British Herbs* he remarks that " He who seeks the herb for its cure, will find it half effected by the walk."

By the sale of his medicines and of his pamphlets relating to medicinal plants, some of which ran through many editions, he made large sums of money.

Before passing on to a consideration of Hill's botanical work brief comment may be made on his literary activities other than those already alluded to. It has already been mentioned that much of his output represented mere hack work, so that it is not surprising to learn, in view of the large amount of work he did, that a certain proportion of it was careless and slovenly, and shewed marked signs of undue haste in production, with the result that his reputation suffered. One work, entitled *Letters from the Inspector to a Lady with the genuine Answers* (1752), is an amorous correspondence not remarkable for its reticence of statement; it reminds one of a similar, but more proper, correspondence, which had a vogue a few years ago.

Hill did not always write for gain, thus *Thoughts concerning God and Nature* (1755) shews him in a different light. This was written from conscientious and religious motives in answer to a book written by Henry St John Viscount Bolingbroke, and was published at a loss, for the number printed, even if all were sold, would not have paid the expenses of production.

His dramatic pieces were of a mediocre nature, and with regard to his novels and other works Baker[1] states that " In some parts of his novels incidents are not disagreeably related, but most of them are nothing more than narratives of private intrigues, containing, throughout, the grossest calumnies, and

---

[1] *Biographica Dramatica.*

aiming at the blackening and undermining the private characters of many respectable and amiable personages. In his essays, which are by much the best of his writings, there is, in general, a liveliness of imagination, and a prettiness in the manner of extending perhaps some very trivial thought; which, at the first *coup-d'œil*, is pleasing enough, and may, with many, be mistaken for it; but, on a nearer examination, the imagined sterling will be found to dwindle down into mere French plate."

In addition to his literary work Hill found time to undertake official duties. In 1760 he was gardener at Kensington Palace, a post which brought him in an income of £2000 per annum[1]; also he was Justice of the Peace for Westminster. According to Mrs Hill[2] he was nominated Superintendent of the Royal Gardens, Kew, and as such he is described on his portrait; his nomination, however, does not appear to have been confirmed, for Thiselton-Dyer[3] states that there is no evidence of his ever having occupied such a position. Hill also advised, at the request of the Earl of Bute, the governors of various islands regarding their cultivation, for which work he received no remuneration.

Anatomy.

Anatomical investigations during the eighteenth century were very barren of results, no real advance upon the discoveries of Grew, Malpighi and others being made. The work of Hill in this field forms no exception to this statement; and, although he accomplished a fair amount of anatomical work, his investigations apparently were without result in the advancement of this particular branch of knowledge.

In 1770 Hill published a small octavo volume on *The*

---

[1] *Dict. Nat. Biog.*

[2] His second wife, the Hon. Henrietta Jones, sister of Charles Viscount Ranelagh. She published " An Address to the Public setting forth the Consequencies of the late Sir John Hill's acquaintance with the Earl of Bute," 1788.

[3] *Historical Account of Kew to 1841, Kew Bulletin,* 1891.

[4] Further information relating to Hill's public Life will be found in the following works. Arthur Murphy, *The Life of David Garrick,* London, 1801; *A narrative of the affair between Mr Brown and the Inspector,* London, 1752; *The Covent Garden Journal,* 1752; Frederick Lawrence, *Life of Fielding,* London, 1855.

Construction of Timber. In order that other investigators might benefit from his experience he fully described and figured the instruments used; of particular interest is a small hand microtome with which he cut his sections. This ingenious tool was the invention of Cummings, and does not differ in essentials markedly from some the writer has seen in use; Hill claims that when the cutter was particularly sharp sections no thicker than a 2000th part of an inch could be obtained. The microscope was made by Adams under the direction of Hill and his patron, unnamed in the book, but in all probability Lord Bute, and embodied some improvements on earlier instruments. This microscope is figured in Carpenter's work on *The Microscope and its Revelations*[1].

The *Construction of Timber* is well arranged: the work begins with a general description of the tissues and their disposition in a thickened stem; then follows a more detailed account of the separate tissues; and finally much space is devoted to a comparison of different tissues in various plants.

Hill's account is fully illustrated with copper plates; his figures of sections are not highly magnified, some not more than twelve times, and their quality is not equal to the best in Grew's *Anatomy*.

Hill principally studied transverse sections, and consequently fell into errors which he might have avoided by the careful observation of longitudinal ones; also he used macerated material, but as his method preserved only the stronger walled elements he did not gain to any great extent from their use.

The parts devoted to comparative anatomy are not at all bad, and they give a concrete idea of the differences obtaining in the different plants.

He apparently understood the nature of the annual rings, and of them he wrote as follows: "These are the several coats of Wood, added from season to season. It has been supposed that each circle is the growth of a year; but a careful attention to the encrease of wood has shewn me, beyond a doubt, that two such are formed each year; the one in the Spring, the other soon after Midsummer." His illustration, however, is not so clear as

---

[1] Ed. by Dallinger, London, 1891.

his statement. Also he realized that the wood vessels were in some way connected with water :

"These vessels arise in the substance of the Wood, principally towards the outer edge of each circle. They are very large in the outermost coat; and smaller in the others: and there are also irregular ranges of them, running thro' the thicknesses of the circles; besides these principal ones of the outer course. They have solid, and firm, coats; and they contain in Spring, and at Midsummer, a limpid liquor, like water, but with a slight acidity: at all other seasons of the year they appear empty, their sides only being moistened with the same acid liquor. Those who examined them at such seasons, thought them air vessels; and in that opinion, formed a construction for them, which Nature does not avow."

Although Hill recognized the entity of the cell he had, in common with his contemporaries, no clear conception of its real nature.

In describing the pith of the rose he does not go astray, and he fully appreciated that the seemingly double contour of the cell walls, when seen in some sections, is due to the thickness of the section with consequent overlapping of the cells; on the other hand he went very wrong in the case of the pith of the walnut, the cavities of which he supposed to be cells like those of the rose, only very much larger and uniseriate as the following quotation shews :

" The Pith of the Walnut consists only of one range of these bladders ['Blebs' or cells], smaller at the edges, largest in the middle, and laid very exactly one upon the other."

When he considers the structure of more or less square or oblong cells his ideas are very wrong. In such cases he thought that the transverse walls were spaces, and the longitudinal walls vessels; curiously enough Hedwig made a similar mistake some years later, possibly he was led astray by Hill's misconception.

Hill adversely criticized the theory that the pith is an organ of propagation, and substituted the view that the corona—i.e. the peri-medullary zone—is all important in this connexion, "From it arises the branches, and encrease of the tree."

Hill had considerable technical ability and, I think, was capable of greatly advancing anatomical botany; unfortunately, however, he gave too little time and thought to his investigations.

Physiology.

The eighteenth century saw the birth of vegetable physiology, Hales and Knight being the two great pioneers in this country. The former flourished in the early part of the century, whilst Knight, although born in 1758, published his great work in 1806.

The chief physiological work of Hill is embodied in a pamphlet of 59 pages, entitled *The Sleep of Plants and Causes of Motion in the Sensitive Plant explain'd*, published in London in 1757, a year previous to the appearance of Du Hamel's *Physique des Arbres*. The paper is in the form of a letter to Linnaeus, and in it the author explains his position with regard to his earlier criticisms of the Linnaean system of classification.

The work is divided into sections, the first of which consists of a brief historical *resumé*, the opinions of Acosta, Alpinus, Ray and Linnaeus on this subject being alluded to. No mention, however, is made of the observations of Bonnet and of Mairan to the effect that the periodic movements of *Mimosa pudica* continued when the plant was kept in prolonged darkness.

In Section 2, after describing the structure of a leaf, Hill remarks that " Leaves are always surrounded by the air; and they are occasionally and variously influenced by heat, light, and moisture. They are naturally complicated, and they act on most occasions together. We are therefore to observe, first, what effects result from their mutual combinations in a state of nature: and having assigned in these cases the effect to the proper and particular ˌcause, from this power of that agent, whichsoever it is, that acts thus in concert with the rest, we may deduce its operations singly."

This passage, although not particularly clear, indicates that Hill fully appreciated the fact that the reaction exhibited by a plant organ is a response to the resultant of a number of forces, and that each factor must be examined separately.

He then goes on to describe his observations on *Abrus*; the structure of the leaf, more especially the course of the vascular bundles, is first dealt with, and then an explanation of the action of light is given. Needless to say, in view of the state of physical science at this period, his explanation, although ingenious, is wide of the mark. He wrote that "Light is subtile, active, and penetrating: by the smallness of its constituent parts, it is capable of entering bodies; and by the violence of its motion, of producing great effects and changes in them. These are not permanent, because those rays which occasion them, are, in that very action, extinguished and lost.

"Bodies may act on light without contact; for the rays may become reflected when they come extreamly near: but light can act on bodies only by contact; and in that contact the rays are lost. The change produced in the position of the leaves of plants by light, is the result of a motion occasioned by its rays among their fibres: to excite this motion, the light must touch those fibres; and where light touches, it adheres, and becomes immediately extinguished....The raising of the lobes in these leaves will be owing to the power of those rays which at any one instance fall upon them: these become extinguished; but others immediately succeed to them, so long as the air in which the plants stands, is enlightened."

Although it was not until 1822, when Dutrochet pointed out the true significance of the pulvini, Hill recognized that these structures were concerned with the movements of the leaflets, not only in the case of *Abrus*, but also in *Mimosa*. He remarked that "It is on the operation of light upon these interwoven clusters of fibres [which are placed at the bases of the main rib, and of the several foot-stalks of the lobes], that the motion of the leaves in gaining their different positions depends; and consequently, the motion itself is various according to the construction of these fibres.

"In the Abrus they are large, and of a lax composition; consequently the lobes are capable of a drooping, an horizontal, and an oblique upward position: in the Tamarind, and the broad-leaved Robinia, they are more compact, and hence all the motion of which those leaves are capable, is an expanding open

and a closing sideways; which the direction and course of the
fibres also' favours: in the Parkinsonia they are smaller, and
yet more compact; and the consequence of this is, that its
lobes have no farther possible motion, than the expanding
upwards."

Again, "The clusters of fibres are as a kind of joints on
which their lobes are capable, under the influence of light,
of a certain limited motion."

Further, with regard to *Mimosa*, he remarks that "To propa-
gate the motion when the leaves are in a state to shew it, there
requires a perfect and confirmed state of those clusters of fibres
lodged at their base." Hill then describes the experiments upon
which he based his conclusions; these shew that he was fully
awake to the importance of keeping the conditions of an experiment, other than those of light, as near constant as possible,
and that the position assumed by the leaves depends upon the
intensity of the light.

His final experiment was to place the *Abrus* in a bookcase
in such a position that the sun shone full upon it; when the
leaves were fully expanded he closed the doors and found that
in an hour "The lobes were all drop't, and it was in the same
state that it would have shewn at midnight. On reopening the
doors the elevated position of the leaves was assumed in twenty
minutes."

Hill offers the same explanation of the movements of *Mimosa*
as of those exhibited by *Abrus*, the reason for their greater
conspicuousness in the former plant being due to the fact that
in *Mimosa* "As there are no less than three sets of these clusters
[of fibres which are placed at the bases of the foot-stalks], the
effects of the same principle are naturally much greater than in
the Abrus where there is only one."

Hill carefully observed the sequence of motion in the *Mimosa*,
and points out that the effect of absolute darkness on the plant
is greater than the rudest touch. He also found that the contact
stimulus must be of a sufficient intensity, and that the degree
of the subsequent motion depended upon the potency of the
stimulus. He further observed that shaking the plant had the
same effect as contact stimulation; also he remarks upon the

fact that the movements of the *Mimosa* and of the Tamarind are less well-marked at a temperature lower than that in which the plants have been reared. Hill considered that "This is probably due to the juices stagnating in the clusters of fibres, and to the contraction of the bark by cold." His explanation of the response to the contact stimulus is of course quite wrong; it may, however, be quoted as an illustration of the view, current at that time, that such motion was due to the fibres which acted like those of muscle. "The vibration of the parts is that which keeps the leaves of the sensitive plant in their expanded and elevated state: this is owing to a delicate motion continued through every fibre of them. When we touch the leaf, we give it another motion more violent than the first: this overcomes the first: the vibration is stopped by the rude shock: and the leaves close, and their foot stalks fall, because that vibrating motion is destroyed, which kept them elevated and expanded.... That the power of motion in the sensitive plant depends upon the effect of light on the expanded surface of the leaves, is certain; for till they are expanded, they have no such power. The young leaves, even when grown to half an inch in length have no motion on the touch, tho' rough and sudden."

Hill fully appreciated the importance of comparative observations; he compared the movements, in response to light, of *Abrus* and *Mimosa*, which plants he placed side by side so that the conditions of the experiment might be the same for each. He found that "In these and in all others, the degree of elevation or expansion in the lobes, is exactly proportional to the quality of the light: and is solely dependent upon it."

Reference also may be made to Hill's views on reproduction[1]; he considered that the pollen grain contained the embryo which was set free by the bursting of the grain after it had been deposited upon the stigma. The stigmatic hairs or papillae were supposed to be the ends of tubes into which the embryos entered, made their way into the placenta, and thus arrived into the "shells of the seeds" (the ovules). It is unnecessary to point out the absurdities of these ideas, but it may be mentioned that Hill's interpretations of his observations were at fault rather

---

[1] *Outlines of a System of Vegetable Generation*, London, 1758.

than the observations themselves. Thus, judging from his
figures, he saw the contents of the pollen grain, the appearance
of which, under the conditions of observation, might easily
suggest the idea of an embryo. Also he noticed that the pollen
grains burst in a little while when placed in water, a phenomenon
which was rediscovered 138 years later[1], and he therefore thought
that a similar bursting, with a consequent setting free of the
embryo, would take place on the wet stigma of the lily, for
example.

Taxonomy.

One of Hill's more interesting works in this branch of Botany
is his *British Herbal*[2]. In it are described a large number of
plants which are illustrated by 75 copper plates engraved by
various artists. None of these plates are of outstanding excellence, indeed many of them are very poor, and their quality is
uneven. Those in the folio consulted by the present writer were
ruined by being coloured.

The plants described are arranged on a system which is not
altogether without interest as it, in a small degree, foreshadows
later systems. It may be indicated by giving the characters of
the first four classes.

Class 1. Plants whose flower consists of several petals, with
numerous threads in the center, and is followed by a
cluster of naked seeds.

Class 2. Plants whose flower consists of several petals, with
numerous threads in the center, and whose seeds are
contained in several pods.

Class 3. Plants whose flower consists of a single petal, and is
succeeded by several capsules.

Class 4. Plants with the flower formed of a single petal, plain,
and of a regular form and succeeded by a single capsule.

It will be seen that Hill relied much on the characters of the
corolla and the gynaeceum. But the chief interest in this work

---

[1] By Lindforss in 1896.

[2] *The British Herbal; an History of Plants and Trees, natives of Britain,
cultivated for use, or raised for Beauty*, London, 1756.

is, perhaps, Hill's criticisms of Linnaeus. One example will suffice ; Linnaeus is criticised for placing *Myosurus* among the pentandria polygynia and thus separating it from *Ranunculus*, *Adonis*, etc. Hill remarked that thus to separate these plants merely because the number of stamens in *Myosurus* is less than in *Ranunculus* is unreasonable since they agree in all other essentials. He himself, however, made a similar error, for it will be observed that in the system followed in the Herbal, *Ranunculus* falls into the first class and *Helleborus* into the second.

These criticisms of Linnaeus, however, are not all of an adverse nature ; in many places Hill does not stint his praise ; and he does not fail, after describing each Genus, to mention its position in the Linnaean System.

Pulteney[1] found it difficult "to reconcile the praises this author bestows on Linnaeus, in many of his writings, with the censures contained in his British Herbal." The difficulty is not very apparent ; Hill sufficiently indicated his position in the following passage taken from the *Sleep of Plants*. "If our opinions have differed, 'tis upon a single Point ; your arrangement of plants. In regard to that much greater article, the establishing their distinctions, and ascertaining their characters, I have always admired and reverenced you : to dispute your determinations there, were to deny the characters of nature.

"Free in the tribute of applause on this head, I have on the other been as open in my censures ; equally uninfluenced by envy, and by fear. It is thus science may be advanced ; and you will permit me to say, thus men of candour should treat one another."

Linnaeus is also criticised in the *Vegetable System*, more particularly for his unnecessary introduction of new names for plants ; but here again Hill is full of praises for Linnaeus's descriptions of species.

Although opposed to the Linnaean system Hill recognised its value as a means of evolving order out of chaos, and to him falls the credit of introducing it into England.

[1] *Historical and Biographical Sketches of the Progress of Botany in England*, London, 1790.

Its first introduction was in his *History of Plants* (1751), but it was unsatisfactory since the *Species Plantarum* was not published until 1753. Hill next explained it in 1758[1], but it was not until two years later that the first *British Flora*, arranged on this system, appeared[2]. According to Pulteney[3], Hill performed this task " in a manner so unworthy of his abilities, that his work can have no claim to the merit of having answered the occasion : and thus the credit of the atchievement fell to the lot of Mr William Hudson F.R.S."

Mention has been made of Hill's *Vegetable System*[4]: a work which consists of 26 folio volumes and was undertaken at the suggestion of Lord Bute. It was commenced in 1759, and the date of the last volume is 1775, the year of Hill's death. No expense was spared in its production, the paper is of the best, and there are 1600 plates : with regard to these the title-page of the work states that they were designed and engraved by the author, but it appears from other sources that they cost four guineas each to engrave, and since it is stated on the auctioneer's announcement of the sale of the copyright (1782), together with some of the original drawings and the remaining sets, that the engravings were made by the best masters under the immediate supervision of the author, it must be concluded that Hill was not the actual engraver although he may have made the original drawings. Attention is drawn to this point, since it casts some doubts as to whether Hill engraved those plates, signed by him, illustrating some of his other works, for instance, *The British Herbal*, and *A Method of Producing Double Flowers from Single*[5], of which some are very good indeed, and, if Hill were the engraver, shew that he had considerable artistic and technical ability.

[1] *The Gardeners New Kalendar...The System of Linnaeus also explained*, London, 1758.

[2] *Flora Britanica sive Synopsis methodica stirpium Britanicarum post tertiam editionem synopseos Raianae...nunc primum ad C. Linnaei methodum disposita*, London, 1760 (some copies are dated 1759).

[3] *Loc. cit.*

[4] *The Vegetable System Or, the Internal Structure and The Life of Plants; Their Parts and Nourishment Explained; Their Classes, Orders, Genera, and Species, Ascertained and Described...*London, 1759—1775.

[5] London, 1758.

Naturally the plates in the *Vegetable System* are of uneven quality, some are very good and not only are pleasing from the artistic point of view, but also give a concrete idea of the plants represented. It is impossible here to criticize this work in detail; but some idea of its scope may be given. The first volume and part of the second is concerned with the history of Botany; the origin of Systematic Botany; the Systems of Caesalpinus, Morison, Ray, Tournefort, Boerhaave, Linnaeus, and others; morphology, anatomy, physiology; and the effect of heat, light, air, soil and water on vegetation. The rest of the work is occupied by descriptions of plants, both British and foreign, when the latter, the native country is mentioned ; in all cases the medicinal properties are given.

It is hardly necessary to remark that notwithstanding the price of the work, 38 guineas plain and 160 guineas coloured, Hill lost considerably over its publication. From Mrs Hill's account[1], it appears that Bute undertook that Hill's circumstances should not be injured by the venture, an undertaking which was not kept; and further, after the death of Hill, Bute refused to compensate Mrs Hill for the unfinished last volume or to take the materials which had accumulated for it out of her hands. Allowing some discount for the natural exaggeration of a bereaved lady suffering from a grievance, there appears but little doubt that the Earl of Bute proved lacking in good faith.

Considered as a systematist there can be no doubt that Hill knew his plants; and although the systematists of the period were overshadowed by Linnaeus, Hill preserved his independence of thought, and did not hesitate to express his opinions when they differed from those of his great contemporary. Although he highly appreciated the work of Linnaeus he disliked his system of classification on account of its artificiality, and he intended to bring forward a natural system of his own. It is not, I think, too much to say that time has justified his criticism; and many of his minor differences have been

[1] *An Address to the Public...loc. cit.*

warranted. For instance, Linnaeus merged the genera *Valerianella* and *Linaria* into those of *Valeriana* and *Antirrhinum* respectively; Hill however recognized the generic rank of the two former[1].

Incidentally, it may be remarked that the acceptance of the year 1753 as the starting-point for the citation of names by the Vienna Botanical Congress has been the cause of more general recognition of Hill's activity in this direction; thus in recent editions of *British Flora* his name is appended to many genera and species[2].

The *Vegetable System* gained Hill the Order of Vasa, from the King of Sweden, in 1774, so that he styled himself Sir John; he was also a Member of the Imperial Academy, and a Fellow of the Royal Academy of Sciences, Bordeaux.

Hill died of gout on the 21st of November, 1775, at about the age of 59, in Golden Square, and was buried at Denham. Notwithstanding the large sums of money he had made, he died heavily in debt owing to the great expense entailed by the publication of the *Vegetable System* and his own personal extravagance. His library was sold in 1776–7, and it has already been mentioned that the copyright of the *Vegetable System* was disposed of by auction.

It is always a matter of difficulty to appraise a man's character, and more particularly is this true of Hill whose character, as Whiston[3] has truly remarked, was so "mixed that none but himself can be his parallel." In the *Sleep of Plants* the following passage occurs: "There is a freedom of style, and assumed manner peculiar to this kind of correspondence, which would be too assuming in works addressed immediately to the public; and might not unnaturally draw upon the author a censure of self-sufficiency and vanity. This explanation, I hope, will defend me from so unfair a charge: for indeed no one knows more the narrow limits of human knowledge; or entertains an humbler opinion of the returns of years of application."

[1] See *Helleborine* Hill v. *Epipactis* Adans. G. Claridge Druce, *Journal of Botany*, XLVI. 1908.

[2] Babington's *Manual of British Botany*, ed. by Groves, London. Hayward's *Botanist's Pocket Book*, ed. by Druce, London, 1909.

[3] John Nichols, *Literary Anecdotes*, 1812.

Nothing could be more proper than this, but against it must be set the opinion of men of his own time, as expressed in the quotation on p. 88, taken from Baker's *Biographica Dramatica.* Many estimates of the character of Hill have been put forward, the first of any authority being that of Johnson[1]:—
"The King then asked him what he thought of Dr Hill. Johnson answered, that he was an ingenious man, but had no veracity; and immediately mentioned, as an instance of it, an assertion of that writer, that he had seen objects magnified to a much greater degree by using three or four microscopes at a time than by using one. 'Now,' added Johnson, 'everyone acquainted with microscopes knows, that the more of them he looks through, the less the object will appear.'...'I now,' said Johnson to his friends, when relating what had passed, 'began to consider that I was depreciating the man in the estimation of his sovereign, and thought it was time for me to say something that might be more favourable.' He added, therefore, that Dr Hill was, notwithstanding, a very curious observer; and if he would have been contented to tell the world no more than he knew, he might have been a very considerable man, and needed not to have recourse to such mean expedients to raise his reputation."

If Hill's reputation for lying rests on no surer foundation than this, he must be held acquitted of much that is charged him. In the above quotation the term microscopes must be read lenses; thus Johnson's reason for his opinion is unfortunate and clearly shews, as Bishop Elrington has remarked, that Johnson was talking of things he knew nothing about. This is the more to be regretted since the opinion of a man of Johnson's rank, who was contemporary with Hill, might have biassed the judgment of smaller and later men.

According to Fitzgerald[2], Hill was a "quack and blustering adventurer," the "Holloway of his day," endowed with "cowardice that seemed a disease." This author is, I think, prejudiced, and his estimate appears to be based upon the least creditable of Hill's performances without giving a proper value to the

[1] Boswell's *Life of Johnson*, ed. by Fitzgerald, London, 1897.
[2] *Ibid.; Life of Garrick, loc. cit.*

better side of his nature and work. On the other hand the
author—a grateful patient—of the short account of the life of
Hill[1] went to the other extreme. This account is entirely
laudatory, and describes Hill as being little short of a genius
surrounded and continually attacked by "envious and malevolent
persons" who "did not fail to make use of every engine male-
volence could invent, to depreciate the character and the works
of a man, whom they saw, with regret, every way so far their
superior."

Disraeli[2] speaks of Hill as the "Cain of Literature," and,
whilst being fully alive to his "egregious egotism" and other
defects of character, he appreciates his worth and recognizes
that Hill was born fifty years too soon. Also he gives him
credit for his moral courage in enduring "with undiminished
spirit the most biting satires, the most wounding epigrams, and
more palpable castigations."

The general concensus of opinion, much of which does not
appear to have been independently arrived at, is that Hill's
nature contained little that was commendable. At the same
time his remarkable industry and versatility were recognised.
His independent and quarrelsome nature, coupled with his mode
of attack and fearlessness in expressing his opinions, made him
cordially hated, and caused much that he did to be viewed with
a prejudiced eye; for instance, it is generally stated that he
obtained his degree of Doctor of Medicine (St Andrews, 1750)
by dishonourable means. Mr Anderson, Librarian and Keeper
of the Records of St Andrews University, has kindly looked the
matter up and informs me that there is nothing whatever to
warrant such a statement; the degree was granted according to
the practice of the time.

It is important to remember that Hill in his earlier days
suffered much from penury, which, to a certain extent, may
have embittered his nature. However this may be, he learnt
subsequently the advantages conferred by a good income, and
was not desirous of becoming reacquainted with his earlier
experiences. This may explain much of his peculiar behaviour.

[1] Edinburgh, 1779, *loc. cit.*
[2] *Loc. cit.*

Disraeli[1] suggests that, in offering himself as Keeper of the Sloane Collection, at the time of its purchase for the British Museum, Hill was merely indulging in an advertisement. Hill probably was sufficiently shrewd to realize that a ready sale for his wares would obtain so long as he kept within the public eye, and much of his extraordinary behaviour in public may have been merely self-advertisement.

The portrait of Hill prefacing this sketch is after Neudramini's engraving of Coates's portrait (1757); the plant represented is a spray of a species of *Hillia*, named in honour of Hill by Jacquin.

[1] *Loc. cit.*

ROBERT BROWN

1773—1858

By J. B. FARMER

Position of Botany before Brown—narrative—diary—naturalist to the
Flinders expedition—travels in Australia—his method in the field—
Essay and Prodromus on the vegetation of New Holland—the Proteaceae and Asclepiadaceae—Brown's digressions—his tenacity and
caution—impregnation—views on the morphology of the Gymnosperms
in the memoir on *Kingia*—foundation of ovular morphology—cell
nucleus discussed—the simple microscope—"Brownian movement"
investigated—summary of other work—Bryophytes—interest in fossil
plants—personal characteristics—Asa Gray's story—the Banksian collections — the British Museum and Linnean Society — contemporary
appreciation—his outstanding merits.

SOMEONE has affirmed that no man is greater than the age
in which he lives. A cryptic utterance, savouring perhaps of a
certain dash of impressionism, and not altogether false as it is
assuredly not wholly true. If, however, we endeavour dispassionately to appraise the performance of the world's great
(though perhaps we should exclude the few greatest) men we
shall probably discover that the implied limitation is justified,
at least in part, by history and experience. The fact is that
hardly anyone can really penetrate far into nature's secret places
without losing his way. The virgin lands of knowledge that lie
beyond the area of contemporary possession are first invaded by
those who can breach the barriers that oppose advance, for genius,
by its wider outlook enables those who are endowed with it to
recognise the weaker spots in these barriers, and thus to lead
the attack. But the new territory, even after it is won, is ever
surrounded by unknown regions, still waiting to be overrun

Plate XI

ROBERT BROWN (*circa* 1856)

when, but not until, the conditions for further expansion shall have been fulfilled.

At the beginning of the nineteenth century the time was ripe for such an addition of new territory to the regions of Botany already occupied at that period. In England, at any rate, the work inaugurated by Ray and others had become overshadowed by the authority of Linnaeus, and even on the Continent the effective advance of the science was for various reasons almost stayed. It is true that in France the Jussieus had started advance on fruitful lines, and others like De Candolle were endeavouring to feel their way through the maze of dimly comprehended relationships, but their efforts were obscured by the growing and fatal facilities for piling up mere catalogues of plants without the clues necessary to direct their energies into more profitable channels. As regards the flowering plants, there was, it is true, a groping after a partially perceived natural system, but the lower ranks of the vegetable kingdom formed, so far as scientific purposes were concerned, a *terra incognita*, and the attempts to elucidate the morphology of these groups in the light of the angiosperms were, as we now can see clearly enough, plainly foredoomed to failure.

Facts were distorted and observations misinterpreted in ways that now seem to us almost to smack of sheer perversity, but we must not forget that the methods which in later years have proved so effective had not then been recognised ; Hofmeister, with his marvellous genius, had not as yet arisen to shew the way through the maze of the lower forms.

But what does strike one as astonishing, or might do so if the circumstance were not still so common, is the evidence of the difficulty men experienced in really seeing things as they were, and of distinguishing the fundamentally important from the trivial or even irrelevant.

As always, what was needed was the man who could fix his gaze on facts, who would spare no pains to find out what was true, and thus succeed in discovering a sure base to serve as a vantage ground for further advance. Von Mohl was one of these, and earlier in the century there was the man, the subject of this lecture, who by his single-hearted search after truth, and the extraordinary ardour and ability with which he prosecuted his investigations will always occupy a high position in the history of Botany.

Robert Brown came of a stock which refused to bow the knee to authority, though his forbears did not, any more than himself, hesitate to impress the weight of it on others. His father was a non-juring clergyman of Montrose, and was in consequence obliged to leave the official ecclesiastical fold. But he carried a congregation with him, and not desiring to set up novel forms of church government, managed to get himself consecrated bishop of the new flock. As bishop, priest and deacon, *tres in uno juncti*, he ministered to his Edinburgh church, and his episcopal staff may still be seen in the rooms of the Linnean Society. His son Robert, who was born in 1773, inherited both his father's independence and also his dominant character. And, indeed, the great influence he wielded in the botanical world was due in no small degree to his strong personality, reinforced as it was by his high scientific attainments.

He began at an early age to evince a love of botany and to give proof of the strong critical faculty which enabled him so successfully to solve the problems he attacked, and so materially to advance our science. He added to his mental attainments a wonderfully methodical habit, and the diary of his earlier years reveals him to us not only as a hard-working student but as one meticulously accurate in detail.

In 1795 he was appointed Surgeon mate to the Fifeshire Regiment of Fencibles, and his letter of appointment signed by the Colonel, James Durham, is preserved in the Natural History Museum. His regiment was quartered in Ireland, and he made good use of his time, collecting all the plants he could get hold of, including mosses and liverworts, of which he amassed a considerable collection. Indeed, it is said that he owed his first acquaintance with Sir Joseph Banks to his discovery in Ireland of the rare moss *Glyphomitrium Daviesii*. This recognition by Sir Joseph proved the turning-point of his life. The six years or so that he spent in the Fencibles were turned to good account, and in looking to his own record of his life during those years one realises how thoroughly he earned the success that crowned his work in after life. There is much humour—perhaps of an unconscious kind, though I am not very sure that it was so very unconscious—in his carefully kept diary. Here is an extract, dated Feb. 7, 1800.

> Before breakfast began the German auxiliary verbs.
> Committed to memory a genus in Cullen's Synopsis. Described Polytrichum aloides—to be compared with Mr Menzies' P. rubellum.
> Began the description of Osmunda pellucida.
> Hospital usual time.
> Took exactly the same walk as on the 4th. Blasia pusilla Lin., Weissia recurvirostra Hedw.? Dicranum varium Hedw., Polytrichum nanum, Polytrichum urnigerum, Phascum subulatum, Dicranum glaucum, absque fruct.
> At dinner about 3 pints of port., remained in the mess room till about 9 or 10 o'clock—slept in my chair till nearly 3 in the morning.
> Feb. 8, before breakfast finished the auxiliary verb *Seyn*, to be, in Wendeborn's German Grammar....

He did not, however, spend all his evenings in this fashion, but whether it was a glass of water, a pint of porter, or what not, it is all gravely set down, together with the work he succeeded in accomplishing. Instances of his thoroughness are not wanting. He says in one place he had read Nicholson's *Chemistry*, ch. vi., on the balance, "to be again perused, my defective knowledge of the mechanical powers rendering part of it unintelligible."

He was fond of reading in bed, but his light literature on these occasions included such works as Adam Smith, Blackstone's *Commentaries*, and a German Grammar.

His botanical acquirements were already attracting notice, and in 1798, being detached for recruiting service, he took the opportunity of a visit to London to utilise the splendid collections in the possession of Sir Joseph Banks, and he was also in the same year elected an Associate of the Linnean Society. Soon after his return to Ireland he received a letter from Sir Joseph offering him the nomination as Naturalist to the *Investigator*, which was to be commanded by Captain Flinders. He at once decided to go, writing, as he tells us, by return of post.

Few men who have, at so early an age, enjoyed the opportunity of a voyage of discovery were so well equipped for the work as was Robert Brown. Blessed with a good constitution, which was also seaworthy, he possessed many physical advantages, but in addition to them he had trained himself as an accurate and accomplished botanist. He spent what time he could spare in London in acquainting himself with all that he could find of the New Holland Flora, and in this connection he had full access to the invaluable Banksian collections.

He was fortunate in having with him on the expedition as draughtsman Ferdinand Bauer, whose beautiful drawings are the admiration of all who know them.

The *Investigator* sailed from Portsmouth in 1801, and on landing at King George's Sound the first collections, amounting to about 500 plants, were made within three weeks. Three days at Lucky Bay yielded 100 species not met with in the previous locality. At Port Jackson the *Investigator* was condemned as unseaworthy, and Captain Flinders determined to return to England to obtain another ship in which to prosecute the expedition. The ship, however, was wrecked in Torres Straits, Brown's duplicate specimens, as well as the live plants on board, being lost, whilst Captain Flinders was held prisoner by the French at Port Louis. Meantime Brown and Bauer continued their travels in Australia, visiting Van Dieman's land as well. Brown subsequently returned to England, oddly enough in the old *Investigator*, in 1805 with a magnificent collection of plants some 4000 in number.

He did not merely *collect*, but he studied his collections on the spot—a method that may be strongly commended to young men who go out as botanists at the present time. His plan was to keep a working herbarium of all the plants gathered by him, as he went along, and he wrote up the descriptions in great part during his actual expeditions. In this way many problems formulated themselves which he was able either to investigate on the spot, or else to lay up additional material for further investigation at leisure. Thus the methodical ways of dealing with the plants collected in earlier years at home stood him in good stead at a time when the opportunities of a lifetime were crowding upon him.

On his return to England he was appointed librarian to the Linnean Society (1805), an office which he held till 1822, and he

at once set about to utilise the vast resources which were now at
his command.

He contributed to the narrative of The Flinders Expedition
an account of the vegetation of New Holland. The essay is
a remarkable one, not only for the masterly descriptions of the
principal genera and orders which it contains, and the critical
remarks which are scattered through the pages, but also for the
geographical and statistical methods of treatment which he
introduced. Many of the orders are new, and Brown shews
his striking perception of affinity not only in his general discussion of the subject as a whole, but also in the definitions of
the new orders and genera which he founded. This soundness of
judgment is shewn on a still larger scale in his more definitely
systematic works such as the *Prodromus*, but one may regard it
generally as an astonishing tribute to his sagacity that very few
of the groups founded by him have needed serious revision,
even when further discoveries made it possible for later botanists
to fill up the lacunae inevitable during those earlier days.

In the year 1810 there appeared the first volume of his great
work, the *Prodromus Florae Novae Hollandiae*. It is a misfortune
that only one volume was ever published, although the work was
advanced in MS. It has been said that a criticism of the author's
Latinity at the hands of a reviewer was the cause of the stoppage
of the publication, but there seems to be no real foundation for
the story. Possibly the expense, coupled with the small return,
may at any rate partly account for it. Be this as it may, Brown
recalled from his bookseller all the unsold copies, and in the
copy preserved at the Natural History Museum there is a list
of the volumes actually sold written by Brown himself, and from
a financial point of view the enterprise clearly proved itself to
be an expensive experiment. The volume as published is
a remarkable work, containing some 450 pages, including
464 genera, nearly one-third of which are here described for
the first time and the number of species amounts to about
2000, some three-quarters of which were new to science. Add
to this the fact that the flora as a whole is very unlike that of
the northern hemisphere, also that the work was accomplished
with such amazing rapidity (largely owing to his particular

methods already alluded to), and one cannot withhold admiration at the energy and the learning of its author. It is a wonderful tribute to his wisdom that his descriptions and arrangements should have so stood the test of 100 years, during which time vast strides in our knowledge of the Australian and other floras have been made. But the lapse of time has resulted in scarcely any but trifling modifications of the general results as he left them. The *Prodromus* is well worth study, for in its pages one constantly meets with hints of observations which have borne fruit in later years. Some of them, indeed, e.g. his observations on Cycads, were expanded by himself into larger treatises in which much light has been thrown on morphological and taxonomic relationships previously but imperfectly understood.

The year before the publication of the *Prodromus*, Brown communicated to the Linnean Society an excellent and learned memoir on the Proteaceae. In this paper we encounter an instance of that whimsical introduction of observations exceedingly valuable in themselves, but mainly irrelevant to the matter in hand, which is a characteristic feature of many of his works. Perhaps it was due to the intense keenness with which he always followed up problems that interested him, so that, like Mr Dick's weakness for King Charles' head, they had to find a place in whatever else he was writing about. Thus his treatise on the Proteaceae starts off with advice to study the flower in the young, instead of only in its adult condition, and this is driven home by an excellent disquisition on the structure of the androecium and gynaeceum of Asclepiads, a subject which occupied his mind for some years and formed the basis for separate papers at subsequent periods. Only when he has discussed the morphology of the Asclepiad flower does he plunge, abruptly, into the questions relating directly to the Proteaceae.

Later on in the same year (1809) he read a masterly paper on the Asclepiadaceae which was subsequently printed in the *Memoirs* of the Wernerian Natural History Society. This Natural Order was here separated by him from the Apocynaceae, from which it had not previously been distinguished, and a correct account of the relations of the remarkable androecium, so characteristic of the Asclepiad flower, was given. Twenty-two years later (in 1831) he again returned to the Asclepiads and described and discussed the mode of pollination and fertilisation in this Order and also in that of the Orchids.

It was characteristic of Brown that he clung with great tenacity to any problem that had once excited his interest. He made himself fully acquainted with the work of his contemporaries and predecessors, and at the same time he constantly attacked it by reiterated first-hand investigations, testing hypotheses and theories by the light of direct observation. He was very cautious, and thus, although he traced the pollen tubes from the pollen grain into the ovary and into the micropyle (foramen) of the ovule, he still leaves it an open question whether, in all cases, anything of a material nature passes from the pollen to the interior of the ovule, which may thus be held responsible for the formation of an embryo.

He cites the observations of Amici and of Du Petit Thouars, and then states he does not feel he is as far advanced as these observers. But in the succeeding pages he traces the tube, of which he says, " the production is a vital action excited in the grain by the application of an external stimulus." We see here a clear perception of the facts of germination and of the operation of what we now call chemiotaxis, for he goes on to add "The appropriate and most powerful stimulus to this action is no doubt contact, at the proper period, with the secretion or surface of the stigma of the same species. Many facts, however, and among others the existence of hybrid plants, prove that this is not the only stimulus capable of producing the effect ; and in Orchideae I have found that the action in the pollen of one species may be excited by the stigma of another belonging to a very different tribe." It is hard to believe that these lines were written so long as 80 years ago. Brown goes on to describe the change that follows impregnation, and the gradual appearance of the embryo. And we must remember that all these observations were made by one who relied almost exclusively on the simple microscope and the simplest—I had almost said barbaric—technique.

He expresses himself in very reserved terms as to the nature

8—2

of the "immediate agent derived from the male organ, or the manner of its application to the ovulum in the production of that series of changes constituting fecundation." But he puts forward the opinion that a more attentive examination of the process in Orchids and Asclepiads is more likely to be fruitful of results than most other families.

He returns again to this matter of fecundation in the following year, studying several orchids, but especially *Bonatea*, for the purpose. He is somewhat shaken as to the validity of his former inferences, and concludes that the "mucous cords" (i.e. strings of pollen tubes) are perhaps derived from pollen "not, however, by mere elongation of the original pollen tubes, but by an increase in their number, in a manner which I do not attempt to explain." In this later paper he also hazards the suggestion that in *Ophrys*, as impregnation is frequently accomplished without the aid of insects, "...it may be conjectured that the remarkable forms of the flowers in this genus are intended to deter, not to attract, insects." Also he suggests that the insect forms in orchidaceous flowers resemble those of the insects belonging to the native country of the plants. This is a clear foreshadowing of what is now called protective mimicry—and the former suggestion is not at any rate wholly without modern supporters, though Brown's share in its origin seems not to be generally recognised.

The keen desire to get to the bottom of a problem, which was so outstanding a feature of Brown's whole mental attitude, unquestionably explains why he was led to make so many important discoveries in such widely different directions. His first hand knowledge of the structure of a vast number of plants gave a soundness and depth to his morphological investigations that must arouse the admiration of everyone who is acquainted with them. He was never satisfied with perfunctory attempts to solve a problem, but, as we have already seen, in the example of his studies on Asclepiads and Orchids, he would return again and again to the matter till he had satisfied himself of the accuracy of his work. It is a pity that all of the present day botanists do not follow more closely in his steps in this respect. Publication of a paper seems to some to be a matter of greater importance than the advance of knowledge by the scientific and scholarly solution of a problem. Such was not Brown's view, and he practised wise delay in publication—*nonumque prematur in annum*, a maxim so strongly advocated by the Latin poet, was really put into practice by him as it also was by some of his contemporaries. Dryander, Solander and others have left, as Brown has done, rich stores of MS. behind them, which have never passed through the press.

The habit of long and continuous reflection on fundamental problems, which was so marked a feature of Brown's character, was perhaps responsible for the curious manner in which some of his most valuable and suggestive contributions to science, and especially to morphology, were given to the world, a habit to which I have already adverted.

We know he had been for many years interested in the ovule, and he made a number of important discoveries respecting it. Closely bound up with this topic were his studies on the Cycads and Conifers. He observed the plurality of embryos in the seeds of these plants, and, indeed, makes a reference to the phenomenon of polyembryony in the *Prodromus*, in which, as in most of his systematic works, morphological observations of the highest value are scattered, though embodied in very compressed phrases, amongst the descriptions of species. But every now and then when writing on one subject he seems to be carried away with the rush of his ideas on general questions. Thus in a memoir on the genus *Kingia* he entitles the paper, possibly to save his face after he had written it, " Character and Description of Kingia ; a new genus of plants found on the south-west coast of New Holland. With observations on the Structure of its unimpregnated Ovulum, and on the female flower of Cycadeae and Coniferae."

This paper is, perhaps, one of the most important of his works, for it was there that, having briefly dismissed the genus *Kingia*, he " let himself go " on the ovule, and then in a masterly dissertation, puts forward his view on the gymnospermic nature of the Cycads and Conifers.

He summarises what was known at that time as to the structure of the ovule, acutely criticising the views of the various

authors he cites.   He emphasises the need of studying the
development in order successfully to interpret the mature structure.   He insists on the origin of the seed coats from the
integuments, on the orientation of the embryo within the amnios
(embryo sac), and on the distinction between the true albumen
which is contained in this "amnios" and the albumen "formed
by a deposition of granular matter in the cells of the nucleus"
(nucellus), i.e. the perisperm, and he goes on to suggest that in
some of these cases the "Membrane of the amnios seems to be
persistent, forming even in the ripe seed a proper coat for the
embryo.... This is the probable explanation of the structure of
true Nympheaceae"...here he seems to have overlooked the
rudimentary endosperm which is really present. Finally he sums
up an admirable account of the whole matter as follows :—" The
albumen, properly so-called, may be formed either by a disposition or secretion of granular matter in the utriculi of the amnios,
or in those of the nucleus itself, or lastly that two substances
having these distinct origins and very different textures may
coexist in the ripe seed as is probably the case in Scitamineae."

He then goes on at once to argue that the apex of the
nucleus is the point of the ovulum where impregnation takes
place, and adds that "all doubt would be removed if cases could
be produced where the ovarium was either altogether wanting
or so imperfectly formed that the ovulum itself became directly
exposed to the action of the pollen or its fovilla."   This leads
him at once to enunciate his view of the gymnospermy of Cycads,
Conifers and Gnetaceae. He reviews very fully the opinions that
had been expressed by others as to the real structure of the
female organ, especially of Pinus, and he mentions the fact that
he himself in the botany of the Flinders' voyage had previously
held the view that a minute perianth was present in the Pine, a
view which, as he says, "On reconsidering the subject in connection with what I had ascertained respecting the vegetable
ovulum" he had now abandoned.

The morphology of the male sporophyll of Cycas, however,
presents a great difficulty, and Brown, less fortunate here,
discusses a number of what seemed to him possible explanations. The recognition of *Sporangia* was remote, and the effort

to homologise the numerous pollen sacs either to grains of pollen which, bursting, liberated fovilla, or to male flowers, or to explain them in other ways, was not very successful. The fact is this was a piece of morphology for which the age was not ready. We must recollect that the comparative morphology of the ovule (in the wide sense) was not attempted, Brown's main contribution to the understanding of this structure consisted in the empirical accuracy with which he elucidated the actual structure—he made no attempt to frame a comparative morphology, for the simple reason that in the condition of knowledge at the time no such comparative morphology was possible or even dreamed of.

Two other remarkable discoveries now demand our attention, and both are instructive as shewing the keenness with which his highly trained powers of observation followed up the clues which his brilliant intellect had enabled him to descry. It was while engaged on a study of the Orchids and Asclepiads that he was led to recognise the existence of the cell nucleus. He worked almost exclusively with what we should call a dissecting microscope. One of his instruments is preserved in the Natural History Museum, and it is well to examine it and reflect on how much may be discerned even with a very primitive instrument if only a good brain lies behind the retina. The "microscope" contains a number of simple lenses of various powers, the highest about $\frac{1}{32}''$ F.L. It is easy with such an instrument to see the nucleus in the epidermal cells when one knows it is there, but to have *discovered* it, and at a time when the technique of staining, &c., was simply non-existent, was a triumph of genius. Brown, of course, could not fully appreciate the great importance of his discovery, but he quite realised that he was dealing with no isolated or trivial fact, and, with characteristic industry and enterprise, he searched many other plants to find out whether his newly recognised nucleus was general or not ; he found it to be so, and we all know how the discovery began at once to bear fruit.

A second observation to which I would refer was also of wide interest, and it was not made merely by chance. Brown was anxious to penetrate if possible into the secrets of fertilisation.

He seems to have been pretty sure that something more than the mere "aura" of older writers was concerned in the matter, and while looking into the evidence for the existence or transmission of material substance, he observed that in the fovilla of the pollen there were vast crowds of minute particles which were in a continual state of dancing motion. He hoped that it might be possible to identify these bodies along their track into the ovule, and so to settle the more urgent questions as to the mode of fertilisation. He states that he made his observations with a simple microscope, the focal length of the lens of which was $\frac{1}{32}''$. Later on he used a much more powerful pocket microscope made by Dollond with power up to $\frac{1}{70}''$ F.L. He got Dollond to check the results with a compound achromatic microscope, and estimated the size of the particles to be $\frac{1}{20,000}$ to $\frac{1}{30,000}''$. Brown was fully aware that he was not the first observer who had seen these moving particles. They had been already noticed by Needham and by Gleichen, but these writers had paid no special attention to them. Brown's great merit in this matter lies in the admirable way in which he conducted the investigation. At first he thought he had lighted upon something which was essentially a peculiarity of the male elements ; then, extending his observations, he had to expand his first idea and admit the "active molecules" to represent a state or condition of living matter generally. As he still further widened the sphere of his investigations, he proved that the same movements occurred in dead tissues, and further that inorganic bodies also exhibited the phenomenon. Later on he found that the movements depended on the minuteness of the particles. He excluded the effect of evaporation, currents and other disturbing influences, and, indeed, the whole investigation shews him to us in the character of an accomplished experimenter as well as a brilliant observer. The complete explanation of these "active molecules," which are in the state generally described as "Brownian movement," still constitutes an unsolved problem, and one finds that it even now continues to occupy the attention of the physicist.

Any attempt adequately to review the whole of Brown's life work is impossible within the limits necessarily imposed by the conditions of a lecture, and I make no pretence to completeness, but will endeavour rather to indicate what appear to be the more important of his many other contributions to science.

His catalogues of the plants collected by those associated with various expeditions, his Kew lists (which were published under Aiton's name) are well known to students of systematic Botany, but his fine monograph on *Rafflesia*, containing, as it does, many observations of general interest will well repay perusal even after these many years. His studies on Cephalotus, on Caulophyllum (with its remarkable seed formation), as well as his considerable memoir on the Proteaceae, shew him as a naturalist imbued with keen insight and possessed of extraordinarily sound judgment.

But Brown did not confine his attention to phanerogams, but, as might have been anticipated from the studies of his earlier years, pursued his investigations into the little explored field of the cryptogams.

We have seen that as a young man he had been greatly attracted to the study of mosses. Later on he contributed two important papers on these plants to the Linnean Society, one in 1809, in which he described two new genera, one of them Dawsonia, the other Leptostomum, both from Australasia. The introductory remarks in which he discusses the character of the moss capsule, are interesting as shewing how hopelessly impossible it was at that time to arrive at a scientific understanding of its structure, so long as everything was tested by the touchstone of the flowering plants. Ten years afterwards he reverted to the same subject, describing the new genus *Lyellia* from Nepaul, and comparing it, as was his wont, with allied genera, e.g. Polytrichum, Buxbaumia and many others, with the view of elucidating the significance of its structure. The spores, however, are still spoken of as seeds. The male plant is generally regarded as the *barren* plant. It is not easy to reconcile the existence of male flowers with the view of Beauvois which Brown seemed still to consider as not disproved, viz. that the seeds and pollen were both contained in the capsule.

Mosses were not the only cryptogams to which he turned his attention. He described a new species of Azolla (*A. pinnata*) from Port Jackson, and the plant was illustrated by the excellent drawings of Bauer. But here, too, the time was not yet ripe for a morphological understanding of the structure. The megasporangium was thought to be the *male* flower, the microsporangia being interpreted as capsules containing several seeds (the glochidia). The explanation of the supposed male flower presented difficulties, but he states that the lower cell (i.e. the megaspore) was once found filled with a powder replacing the turbid fluid ordinarily occurring there, and the powder was supposed in some way to be ejected and thence to be conveyed to the female organ.

Ferns also claimed his attention, and among his other contributions he founded the genus *Woodsia*, calling attention to the character of the involucrum (indusium), which separated it from the other polypodia with which the species had previously been associated.

Brown had always taken a keen interest in fossil plants, although, so far as I am aware, he only wrote one paper on the subject. This one, however, was of considerable importance, for its subject was the Brownian cone of Lepidodendron, called by him Triplosporite, though its true affinities were correctly gauged.

Although, as I have said, Brown was less successful when grappling with cryptogams, he is always worth reading on any subject, and in his own special province, that of the flowering plants, I know of no one amongst the older writers from whom one may learn so much. This is due not only to the genius and erudition which he brought to bear on every problem he attacked, but also to the example he affords of scientific method in handling his subject. In his respect for accuracy, in his cautious attitude, as well as in the single-minded honesty of purpose he everywhere exhibits, he has set an example not only for his own but for all future time.

His personal character made a deep impression on his contemporaries. To his friends he was very faithful, and the unanimous tribute of affectionate (though respectful) admiration affords full proof of this. Like many other strong characters, however, he seems also to have been able at times to shew a

rougher side of his nature. He was not generous with his
specimens, nor was he always ready to part with information.
Asa Gray tells a story of how he encountered this trait of
Brown's character. Gray was visiting this country and, of
course, made the great botanist's acquaintance. One day Brown
told him that he knew of a character by which *Rhexia* (a genus
in which Gray was at that time interested) could be distinguished
from some nearly allied ones, and that this character had escaped
the notice of De Candolle and others. But Gray could not get
it out of him, and it was not till the following week that Brown
was induced to part with his secret !

It is interesting to observe the impression the elder botanist
made on Gray, and to note the growing admiration with which
the younger man speaks of him in the very readable diary he
kept of his London visit. It was the same, however, with all. The
more intimate the acquaintance the more profound the respect,
and sometimes the love, that Brown's personality inspired.

Brown was a keen business man, and well lived up to the
traditions of the land of his birth. He gave a remarkable proof
of his canniness in the successful outcome of his bargaining with
the trustees of the British Museum. Sir Joseph Banks by his
will had left him not only his house, but also a life user of the
Banksian collections, after which they were to go to the Museum.
In 1827 Brown entered into a hard agreement with the trustees
to transfer the collections at once to the Museum, he being
appointed "under-librarian" at an adequate salary, with a well
safeguarded position. He used commonly to take 11 weeks'
holiday—a length of vacation which served to differentiate him
rather clearly (and to his own advantage) from his colleagues.
He successfully countered all official moves designed to encroach
on the terms of his agreement whereby his freedom might be
curtailed, and his conditions of service be brought more into
line with those that obtained elsewhere in the Museum.

He maintained through his life intimate relations with the
Linnean Society. He acted during his earlier life as Librarian
to the Society, an office which he resigned in 1822. Two years
previously he had succeeded to the house in Soho Square which
had been left to him by Sir Joseph Banks, and as it was larger

than his own requirements demanded, an arrangement was made by which the Linnean Society moved into the vacant rooms, where it remained for a number of years. Brown subsequently became President of the Society (in 1849).

Robert Brown was deservedly acclaimed by his contemporaries as the first botanist of his age, and honours fell to his share even in his earlier years. He was elected a Fellow of the Royal Society in 1811, and twenty-eight years afterwards was awarded the Copley Medal. He was approached in 1819 in connection with the Chair of Botany in Edinburgh, but decided not to sever his intimate connection with Sir Joseph Banks. Abroad he was probably more widely known than in this country, for when on a visit to Prussia the King sent a special carriage to meet him, and decorated him with the Order *Pour la Mérite*. In England, on the other hand, though held in the highest esteem by his scientific *confrères*, he shared the obscurity that was the common lot of many of the *savants* of that age. He was, however, awarded a civil pension, although not without question on the part of certain members of the House of Commons.

He lived to a ripe age, passing away in the year 1858, the 85th of his age. To the last he retained his interest in his life work, and on June 3, a week before he died, he signed a certificate in favour of an Associate of the Linnean Society.

Robert Brown, as we have seen, penetrated more deeply than most of his contemporaries into the secrets of nature, and he enriched the science to which he devoted his long life by discoveries of fundamental importance. But he, no more than others, was able to anticipate, with all his insight, the recognition of the broader bonds of coherence which link up the plant kingdom as a whole. That was only made possible when the researches of Hofmeister, the great Tübingen Professor, had been made known to the world. But it is no reproach to his memory or to his reputation that he should have fallen into error when attempting to elucidate the critical stages in the life history of cryptogams. The historical interest attaching to his mistakes lies in their inevitableness at the time when he was actively working.

It would be as ungracious as it would be futile to attempt to rob the great botanist of the meed of praise which by all that is right belongs to him, because he could not escape from the influence of limiting factors. His supreme merit rests in his wonderful elucidation of the morphology and inter-relationships of the higher plants, and if we judge him by his achievements in this field we shall hardly disagree with v. Humboldt in according to him the title of *Facile Botanicorum princeps, Britanniae gloria et ornamentum.*

SIR WILLIAM HOOKER
1785—1865

By F. O. BOWER

Early pursuits—appointed to Glasgow—Garden administration—teaching methods — appointed Director of Kew—state of Botany — vigorous development of Kew—serial publications—floristic work—descriptive work on Ferns—his record.

"POETA nascitur non fit." A poet is born, not made. If this be true of poets, much more is it true of botanists. The man who takes up botany merely as a means of making a livelihood, rarely possesses that true spirit of the naturalist which is essential for the highest success in the Science. It is the boys who are touched with the love of organic Nature from their earliest years, who grub about hedgerows and woods, and by a sort of second sight appear to know instinctively, as personal friends, the things of the open country, who provide the material from which our little band of workers may best be recruited.

Such a boy was Sir William Hooker, the subject of this lecture. He was born in 1785, at Norwich. There is no detailed history of his boyhood, but it is known that in his school days he interested himself in entomology, in drawing, and in reading books of natural history, a rather unusual thing at the time of the Napoleonic wars! In 1805, when he was at the age of 20, he discovered a species new to Britain, in *Buxbaumia aphylla*, and his correspondence about it with Dawson Turner shows that he was already well versed not only in the flowering plants, but also in the Mosses, Hepaticae, Lichens, and fresh-water Algae of Norfolk, his native county. Three years later Sir James Smith dedicated to him the new genus

Plate XII

WILLIAM JACKSON HOOKER (1834)

Hookeria, styling him as "a most assiduous and intelligent botanist, already well known by his interesting discovery of *Buxbaumia aphylla*, as well as by his scientific drawings of Fuci for Mr Turner's work: and likely to be far more distinguished by his illustrations of the difficult genus *Jungermannia*, to which he has given particular attention" (*Trans. Linn. Soc.* IX. 275). Clearly young Hooker was a convinced naturalist in his early years, and that by inner impulse rather than by the mere force of circumstances.

Not that the circumstances of his early years were in any way against his scientific tastes. He inherited a competence at the early age of four, and so was saved the mere struggle of bread-winning. His father was personally interested in gardening, while from his mother's side he inherited a taste for drawing. Moreover, he was early thrown into relations with some of the leading naturalists of his time, chiefly it appears by his own initiative, and doubtless he owed much in those opening years to the advice and stimulus of such men as Dawson Turner, and Sir James Smith. Elected to the Linnean Society in 1806, he became acquainted in the same year with Sir Joseph Banks, Robert Brown, and other leading naturalists. Thus when other young men would be feeling for their first footing, he at the age of 21 had already penetrated into the innermost circle of the Science of the country. For a period of sixty years he held there a place unique in its activity. He shared with Augustin Pyrame De Candolle and with Robert Brown the position of greatest prominence among systematists, during the time which Sachs has described as that of "the Development of the Natural System under the Dogma of Constancy of Species." The interval between the death of Linnaeus and the publication of the *Origin of Species* can show no greater triumvirate of botanists than these, working each in his own way, but simultaneously.

The active life of Sir William Hooker divides itself naturally into two main periods, during which he held two of the most responsible official posts in the country, viz. the Regius Chair of Botany in Glasgow and the Directorship of the Royal Gardens at Kew. We may pass over with but brief notice the years from 1806 to 1820, which preceded his attainment of professorial rank. Notwithstanding that notable work was done by him in those years, the period was essentially preparatory and provisional, and can hardly be reckoned as an integral part of his official life. He was in point of fact an enthusiastic amateur, one of that class which has always been a brilliant ornament of the Botany of this country, and has contributed to its best work. He travelled, making successive tours in Scotland and the Isles, no slight undertaking in those days (1807, 1808). In 1809 he made his celebrated voyage to Iceland, described in his *Journal*, published in 1811. But his collections from Iceland were entirely lost by fire on the return voyage. His son remarks that the loss to science was probably greatest in respect of the Cryptogamic collections; this naturally followed from the fact that already he had taken a prominent place as a student of the lower forms, and the field for their study was more open than among the flowering plants of the island. It was among the Cryptogams that Sir William found the theme of his first great work, the *British Jungermanniae*, published in 1816. Nearly a century after its appearance it still stands notable not only for the beauty of the analytical plates, but as a foundation for reference. It must still be consulted by all who work critically upon the group, subdivided today, but comprehended then in the single genus *Jungermannia*. During this period he also produced the *Musci Exotici*, with figures of 176 new species from various quarters of the globe. Thus up to 1820 his chief successes lay in the sphere of Cryptogamic Botany.

Naturally so ardent a botanist desired to widen his experience by travel. But circumstances checked the projects which he successively formed to visit Ceylon and Java, South Africa, and Brazil. In 1814 he went to France, and became acquainted with the leading botanists of Paris. He proceeded to Switzerland and Lombardy, returning in 1815, in which year he married the eldest daughter of his friend Mr Dawson Turner. Meanwhile, at his father-in-law's suggestion he had embarked in a business for which he was not specially fitted by experience or by inclination. It did not prove a success, and as the years drew on, having a young family dependent upon him, he began to look out for some botanical appointment which should at once satisfy his personal tastes, and be remunerative. The chair in Glasgow becoming vacant in 1820 by the transfer of Dr Graham to Edinburgh, he received the appointment from the Crown, largely through the influence of Sir Joseph Banks. He entered upon its duties never having lectured before to a class of students, nor even heard such lectures, but otherwise equipped for their performance in a way that would bear comparison with any of the professors of his time.

Glasgow was in 1820 at an interesting juncture in its botanical history. Though the science of botany had been taught for a whole century in the University, a separate chair had been founded by the Crown only two years before. Moreover, though there had been for a long period a "Physic Garden" in the grounds of the old College, this had proved insufficient, and its position within the growing town unsuitable. Accordingly, in part by grant from the Crown, partly from the funds of the University, but largely by the subscriptions of enthusiastic citizens a Botanic Garden had been founded under Royal Charter in 1817, and opened to the public in 1819. The first blush of novelty had not worn off this new enterprise when a man, already in a leading position, whose successful achievements had shown his quality, acquainted with many of the leading botanists of Europe, and with youth and unbounded energy at his disposal entered upon the scene, and began that course of organisation of Public Botanic Gardens which he continued to the day of his death.

There was nothing to prevent the Glasgow establishment from rapidly taking a leading position. Largely as the result of Hooker's influence and initiative, and assisted greatly no doubt by the zeal with which the movement was supported by individual citizens, and aided by the position of Glasgow as a great commercial centre, contributions to the garden began to come in from every quarter of the globe. Taking the number of species represented as a measure, the growth of the living collections was rapid beyond precedent. In 1821 the number of species living in the garden was about 9000: in 1825 it is quoted at 12,000, while the increase in number from that period onwards was about 300 to 500 per annum. Of these a large number were new species, not previously described or figured. This work Hooker carried out, and the publication of his results widened still further the desire of the officials of other gardens to effect exchanges. In 1828, after it had been in existence but ten years, the Glasgow garden was corresponding as an equal with 12 British and Irish, 21 European, and 5 Tropical gardens, while it had established relations with upwards of 300 private gardens. In 1825 Sir William Hooker published a list of the living plants in pamphlet form, with a plan of the garden, copies of which are still extant. But the following years, from 1825 to 1840 were the most notable in its history as a scientific institution. It is recorded in the minute books that scientific visitors almost invariably expressed the opinion that the garden would not suffer by comparison with any other similar establishment in Europe. It can hardly have come as a surprise to those who had witnessed his work in Glasgow that when a Director had to be appointed to the Royal Gardens at Kew, the post was offered to Hooker. He accepted the appointment and left Glasgow in 1841.

His conduct of the Glasgow professorship from 1820 to 1841 was a success from the first, notwithstanding his entire want of prior experience of such duties. Sir Joseph Hooker, in his speech at the opening of the New Botanical Buildings in Glasgow University, in 1901, pointed out how he "had resources that enabled him to overcome all obstacles: familiarity with his subject, devotion to its study, energy, eloquence, a commanding presence, with urbanity of manners, and above all the art of making the student love the science he taught." Not only students in medicine, for whom the course was primarily designed, attended the lectures, but private citizens, and even officers from the barracks.

Sir Joseph describes his father's course as opening with a few introductory lectures on the history of botany, and the general character of plant-life. As a rule the first half of each hour was occupied with lecturing on organography, morphology, and classification, and the second half with the analysis in the class-room of specimens supplied to the pupils, the most studious of whom took these home for further examination. An interesting event in these half-hours was the professor calling upon such students as volunteered for being examined, to demonstrate the structure of a plant or fruit placed in the hands of the whole class for this purpose. The lectures were illustrated by blackboard drawings, probably these were a special feature in the hands of so experienced an artist as he, and also by large coloured drawings, chiefly of medicinal plants, which were hung on the walls. Another feature, which happily still survives, was the collection of lithographed illustrations of the organs of plants, a copy of which was placed before every two students. The first edition of these drawings appears to have been by his own hand. But in 1837 a thin quarto volume of *Botanical Illustrations* was produced, "being a series of above a thousand figures, selected from the best sources, designed to explain the terms employed in a course of Lectures on Botany." The plates were executed by Walter Fitch, who was originally a pattern-drawer in a calico-printing establishment, and entered the service of Sir William in 1834. This great botanical artist continued to assist Sir William till the death of the latter, and himself died at Kew in 1892. A number of copies of this early work of Fitch remain to the present day in the Botanical Department in Glasgow.

Other branches, however, besides Descriptive Organography were taken up. Naturally the plants of medicinal value figured largely in the course, which was primarily for medical students. Illustrative specimens, of which Sir William gathered a large collection, were handed round for inspection. These, together with other objects of economic interest finally made their way to Kew, and were embodied in the great collections of the Kew Museums. The branch of anatomy of the plant-tissues was not neglected. Of this he wrote at the time of taking up the duties of the chair, "it is a subject to which I have never attended, and authors are so much at variance as to their opinions, and on the facts too, that I really do not know whom to follow." He continues with a remark which is singularly like what one might have heard in the early seventies, just before the revival of the laboratory study of plants in this country. He remarked that "Mirbel has seen what nobody else can: so nobody contradicts him, though many won't believe him." I can hardly doubt that physiology of plants will also have figured in the course, first because Sir William was himself a successful gardener, but secondly because we have in the Botanical Department in Glasgow the syllabus of the lectures of Professor Hamilton who taught botany in the University in the latter end of the 18th century. In this course physiology took a surprisingly large place, and we can hardly believe that it would have dropped out of Sir William's course altogether. But of this there is no definite record.

Another feature of the teaching of Sir William was the practical illustration of botany in the field, by means of excursions. Of these Sir Joseph tells us there were habitually three in each summer session, two of them on Saturdays, to favourable points in the neighbourhood of Glasgow; but the third, which took place about the end of June, was a larger undertaking. With a party of some thirty students, and occasional scientific visitors from elsewhere, he started for the Western Highlands, usually the Breadalbane range. In those days, before railways, and often with indifferent roads, this was no light affair, and in some cases it involved camping. I do not know whether this was the beginning of those class excursions which have been so marked a feature in the botanical work of the Scottish Universities, but it is to be remembered that his immediate successor in the Glasgow chair was Dr Hutton Balfour, who in later years confirmed and extended the practice, and it has been kept up continuously in the Scottish universities ever since. It was to meet the requirements of such work in the field that Sir William prepared and published the *Flora Scotica*. The first edition appeared before his second year's class had assembled in 1821. The first Part related to the Phanerogams only, arranged according to the Linnaean system. The second, which seems to have been almost as much a new book as a second edition, contained the Phanerogams arranged according to the natural system, just then coming into general use. It also embodied the Cryptogams, in the working up of which he had the assistance of Lindley and of Greville. The total number of species described was 1784, of which 902 were Cryptogams.

And thus was initiated that profuse and rapid course of publication which characterised the period of office of Sir William Hooker in Glasgow. The duties of the chair were comparatively light, and only in his later years did he extend them voluntarily into the winter months. He worked year in year out, early and late, at his writing, and rarely left home. The 21 years of his professorship were perhaps the most prolific period of his literary production. It was brought to a close in 1841, by his appointment to the directorship of the Royal Gardens at Kew, which had in March 1840 been transferred from the Crown, under the Lord Steward's Department, to the Commissioners of Woods and Forests. Sir William had been for some time desirous of changing the scene of his activities from the relatively remote city of Glasgow to some more central point, and the opening at Kew not only satisfied this wish, but also put him in command of the establishment in which he saw, even in its then undeveloped state, the possibility of expansion into a botanical centre worthy of the nation.

In the spring of 1841 Sir William removed to Kew, taking with him his library, his private museum and herbarium. This was the first of those incidents of denudation of the botanical department in Glasgow, the direct result of the system that held its place in the Scottish Universities till the Act of 1889. Till that date the chair was "farmed" by the professor. Almost all the illustrative collections and books of reference were his private property. Whenever, as has repeatedly been the case in Glasgow, the occupant of the chair was promoted elsewhere, he naturally took his property with him, and the University was denuded, almost to blank walls. Fortunately that is so no longer. But in the present case the collections were removed, and finally formed the basis of the great museums, and of the herbarium of Kew.

At the time of Sir William's appointment Kew itself was in a very unsatisfactory state. The acreage of the garden was small compared with what it now is. The houses were old, and of patterns which have long become obsolete. Only two of them are now standing, viz. the Aroid house near the great gates, and the old Orangery, now used as a museum for timbers.

There was no library, and no herbarium. In fact Kew in 1841 was simply an appanage to a palace, where a more than usually extensive collection of living plants were grown. In the course of the negotiations which led up to the transfer to the Department of Woods and Forests it had even been suggested that the collections themselves should be parted with. It was to such an establishment, with everything to make, and little indeed to make it from, that Sir William Hooker came at the age of 55. He had, however, unbounded enthusiasm, and confidence in the public spirit, and in himself: and what was still more to the point, the experience gained in the smaller field of Glasgow, in building up the garden there, combined with a knowledge of plants which was almost unrivalled, and acquaintance with the leading botanists and horticulturalists of Europe. It was then no matter for surprise that he should accept the position, even though the initial salary was small, and no official house was provided.

As the date of Sir William's appointment may be said to be the birth-day of the new development of Kew, it will be well to pause a moment and consider the position of botanical affairs in Europe at that time. The glamour of the Linnaean period had faded, and the Natural System of Classification of Plants initiated by De Jussieu had fully established its position, and had been worked into detail, taking its most elaborate form in the *Prodromus Systematis Naturalis* of Augustin Pyrame De Candolle. That great luminary of Geneva died in this very year of 1841, leaving his work, initiated but far indeed from completion, in the hands of his son Alphonse. In England, Robert Brown was in the full plenitude of his powers, and in possession of the Banksian herbarium was evolving out of its rich materials new principles of classification, and fresh morphological comparisons. In fact morphology was at this time being differentiated from mere systematic as a separate discipline. Nothing contributed more effectively to this than the publication of *Die Botanik als inductive Wissenschaft*, by Schleiden, the first edition of which appeared in 1842: for in it development and embryology were for the first time indicated as the foundation of all insight into morphology. But notwithstanding the great advances of this period in tracing natural affinities, and in the pursuit of morphological comparison, branches which would seem to provide the true basis for some theory of Descent, the Dogma of Constancy of Species still reigned. It was to continue yet for 20 years, and the most active part of the life of the first Director of Kew was spent under its influence.

Meanwhile great advances had been made also in the knowledge of the mature framework of cell-membrane in plants. Anatomy initiated in Great Britain in the publications of Hooke, Grew, and Malpighi, had developed in the hands of many "phytotomists," the series culminating in the work of Von Mohl. But it was chiefly the mere skeleton which was the subject of their interest. Eight years previously, it is true (1833), Robert Brown had described and figured the nucleus of the cell, and approached even the focal point of its interest, viz. in its relation to reproduction. But the demonstration of the cytoplasm in which it was embedded was yet to come. In fact, the knowledge of structure omitted as yet any details of that body which we now hold to be the "physical basis of life."

The period immediately succeeding 1841, was, however, a time pregnant with new developments. The study of protoplasm soon engaged the attention of Von Mohl. Apical growth was investigated by Naegeli and Leitgeb. The discovery of the sexuality of ferns, and the completion of the life-story by Bischoff, Naegeli, and Suminski led up to the great generalisation of Hofmeister. And thus the years following 1841 witnessed the initiation of morphology in its modern development. On the other hand, Lyell's *Principles of Geology* had appeared and obtained wide acceptance. Darwin himself was freshly back from the Voyage of the "Beagle," while Sir Joseph Hooker, then a young medical man, was at that very time away with Ross on his Antarctic voyage, and shortly afterwards started on his great journey to the Himalaya. These three great figures, the fore-runner of Evolution, the author of the *Origin of Species*, and Darwin's first adherent among biologists, were thus in their various ways working towards that generalisation which was so soon to revolutionise the science of which Kew was to become the official British centre. Well may we then regard

this date, and the event which it carried with it, as a nodal point
in the history of botany not only in this country, but also in the
world at large.

The urgent necessity for such an official centre as Kew
now is was patent in the interests of the British Empire. The
need of it had already been clearly before the minds of the
Parliamentary Commission, appointed a few years before, with
Dr Lindley as chairman, to report upon the question of the
retaining of the Botanic Gardens at Kew. The report contained
the following passage which, while it formulates an ideal then
to be aimed at, summarises in great measure the activities of
the present establishment at Kew. "The wealthiest and most
civilised country in Europe offers the only European example
of the want of one of the first proofs of wealth and civilisation.
There are many gardens in the British colonies and dependencies, as Calcutta, Bombay, Saharunpore, the Mauritius, Sydney
and Trinidad, costing many thousands a year: their utility is
much diminished by the want of some system under which they
can be regulated and controlled. There is no unity of purpose
among them; their objects are unsettled, and their powers wasted
from not receiving a proper direction: they afford no aid to each
other, and it is to be feared, but little to the countries where they
are established: and yet they are capable of conferring very
important benefits on commerce, and of conducing essentially to
colonial prosperity. A National Botanic Garden would be the
centre around which all these lesser establishments should be
arranged: they should all be placed under the control of the
chief of that garden, acting with him, and through him with
each other, recording constantly their proceedings, explaining
their wants, receiving supplies, and aiding the mother country in
everything useful in the vegetable kingdom: medicine, commerce,
agriculture, horticulture, and many branches of manufacture
would derive considerable advantage from the establishment of
such a system....From a garden of this kind Government could
always obtain authentic and official information upon points
connected with the establishment of new Colonies: it would
afford the plants required on these occasions, without its being
necessary, as now, to apply to the officers of private establish-

ments for advice and help....Such a garden would be the great source of new and valuable plants to be introduced and dispersed through this country, and a powerful means of increasing the pleasures of those who already possess gardens : while, what is far more important, it would undoubtedly become an efficient instrument in refining the taste, increasing the knowledge, and augmenting the rational pleasures of that important class of society, to provide for whose instruction is so great and wise an object of the present administration."

Such were the surrounding conditions, and such the aims of Sir William Hooker when he took up the duties of Director of the Royal Gardens. He was, however, given no specific instructions on entering office. He therefore determined to follow the suggestions of Dr Lindley's Report, and in the carrying of them out he had powerful support, both official and other. The original area of the Garden, apart from the Pleasure Grounds and the Deer Park, was small; when first taken over from the Lord Steward's Department by the Commissioners of Woods and Forests, it extended only to about 18 acres, and the Chief Commissioner, Lord Duncannon, was strongly opposed to their enlargement, or to further expenditure upon them. It required methods of diplomacy, as well as determination and energy, not always to be found among scientific men, to carry into effect the scheme laid down in the Report, and success came only slowly. In 1842 additional ground was taken in from the Pleasure Grounds, so as to afford an entrance from Kew Green, now the principal gate of the Garden. In 1843 there were added 48 acres of Arboretum, including the site of the Great Palm House. This was commenced in 1844, and was followed in 1846 by the Orchid House. In 1848 the old storehouse for fruit (close to the fruit garden of the old Palace, now the site of the Herbaceous Ground), was converted into a Museum of Economic Botany, the first of its kind to be established. It was in part furnished by the collections which Sir William had brought with him from Glasgow. It now stands as Museum No. II. In 1850 the Water-Lily House was built, and in 1855 the long house for Succulents. Meanwhile, in 1853, an official house had been found for the Director, while another Crown house adjoining

Kew Green was handed over for the growing herbarium and library. These, which were in the main if not indeed altogether the private property of the Director, had up to this time been housed in his private residence. Now they found more convenient accommodation, where they would be more accessible for reference, in a building belonging to the establishment. In 1857 the Museum No. I. was opened. For long the collections had exceeded the space in the older Economic Museum (No. II.). This was, however, retained for the specimens belonging to the Monocotyledons and Cryptogams, while those of the Dicotyledons were arranged in the new and spacious building of No. I. In 1861 a reading-room and lecture-room for gardeners was opened, and in 1862 the central portion of the great range of the Temperate House was completed from plans approved in 1859. The wings which now complete the original design were added many years afterwards. In 1863 the old Orangery was disused as a plant-house, and diverted to the purpose of a Museum for Timbers, chiefly of colonial origin. It is now known as Museum No. III. The above may serve as a summary of the more important material additions to the Kew establishment, made during the life of Sir William Hooker. It will be clear that his activity must have been unceasing, in working towards the ideal sketched in the report of Dr Lindley. His efforts never abated till his death in 1865, in the 81st year of his age. The establishment of Kew has developed further as years went on. But as he left it, the essentials were already present which should constitute a great Imperial Garden. Truly Sir William Hooker may be said to have been the maker of Kew, if regard be taken merely of the material establishment.

In no less degree may he be held to have been the maker of Kew in respect of its scientific collections, its methods, and its achievements. To these his own untiring activity contributed the driving force, while his wide knowledge, and ready apprehension of fact gave the broad foundation necessary for successful action. But as the period of development of Kew in these respects was but the culmination of the work already initiated in Glasgow, it will be well to review Sir William Hooker's scientific achievements over the whole of his professional career, including the Glasgow period together with his later years at Kew.

Taking first the living collections, he had already shown at Glasgow, where the opportunities were more limited than at Kew, a singular success in securing additions to the plants under cultivation. This is now reflected more clearly in the lists which were published from time to time than in any actual specimens still living after the vicissitudes of cultivation of 70 years ; though it is not improbable that some of our older specimens date from his period of office. The current floristic serials, many of them produced and even personally illustrated by himself, also form a record of the novelties from time to time secured. This rapid growth of the Glasgow garden has already been noted, and the large number of the plants introduced under his influence. It only required the same methods to be put in practice in the larger sphere of action of the metropolis to ensure a similar, though a far greater result at Kew. Moreover, the official position which he there held as Director, gave an increasing obligation to meet his wishes on the part of foreign and colonial gardens, and other sources of supply. Notable among the many òther living collections that resulted was the series of Ferns, already a subject of his detailed study while at Glasgow. In its maintenance and increase he was ably assisted by the Curator, Mr John Smith, himself no small a contributor to the systematic treatment of the Ferns. Hooker's aim was, however, not to forward the interests of any special group of plants, but to make the collections as representative as possible. This is clearly reflected in the various character of the plant-houses successively built at his instigation, and remaining still to testify to the catholicity of his views.

In the days at Glasgow, Sir William had already made his private museum ancillary to the living collections, in his endeavour to demonstrate the characters of the vegetable world. This line of demonstration he further developed after his removal to Kew, and the results, together with later additions, but with methods little changed, are to be seen in the splendid museums of the Gardens at the present time. The specimens were from the first mainly illustrative of Economic Botany, such as are of service to the merchant, the manufacturer, the dyer, the chemist and druggist, and the physician: or to artificers in wood and in textiles. But the interests of the scientific botanist were not forgotten, while a special feature from the first was the portrait gallery of the leaders in the subject. Thus the museums which he initiated, and were indeed the first Museums of Economic Botany ever formed, are now not the least interesting and certainly among the most instructive features of Kew.

But the centre of the Garden for reference and for detailed study is now the herbarium and library, housed in the large building near to the entrance from Kew Green. To those familiar with that magnificent mine of accumulated learning as it now stands, it may be a surprise to hear that it has grown in the course of less than 60 years out of the private collections of Sir William Hooker, and of his friend Bentham. The story of it may be gathered from the sketch of the Life and Labours of the First Director, published by Sir Joseph Hooker in the *Annals of Botany* in 1903, a work to which I have been largely indebted for the materials for this lecture. The Hookerian herbarium and library were already extensive before it was removed from Glasgow. When the new Director of Kew took up his appointment, neither books nor a herbarium were provided for him: but he was well equipped with those of his own. They were at first lodged in his private house, till in 1853 he moved into the official residence. But the latter did not afford the accommodation for them which the Government had guaranteed. They were therefore placed in a building adjacent to the Botanic Garden. It was further agreed, on condition that the herbarium and library should be accessible to botanists, that he should be provided with a scientific herbarium Curator. Four years afterwards the Royal Gardens came into possession, by gift, of the very extensive library and herbarium of G. Bentham, Esq., which was second only to Hooker's own in extent, methodical arrangement, and nomenclature; and it was placed in the same building. The two collections in considerable degree overlapped, being derived from the same sources. But one great difference between them was that Bentham confined his herbarium to flowering plants, while Hooker's rapidly grew to be the richest in the world in both flowering and flowerless plants. Finally after his death it was acquired by purchase for the State in 1866, together with about 1000 volumes from his library, and a unique collection of botanical drawings, maps, MSS., portraits of botanists, and letters from botanical correspondents, which amounted to about 27,000. These were the prime foundations of the great herbarium and library now at Kew. Great additions have since been made by purchase and by gift, and the building has been repeatedly extended to receive the growing mass of material. But for all time the character and individuality of the collections will remain stamped by the personality of those two great benefactors, Bentham and the first Hooker.

Sufficient has now been said to indicate that Hooker's work was that of a pioneer, in providing the material foundation necessary for the further study of the science, not only in this country, but also in the furthest lands of the Empire. He supplied a coordinating centre for botanical organisation in Britain, and for that service he has earned the lasting gratitude of botanists. It remains to review his own published works, and base upon them some estimate of his more direct influence upon the progress of the science. We shall see that in this also his work was largely of that nature which affords a basis for future development. It was carried out almost entirely under pre-Darwinian conditions. He was pre-eminently a descriptive botanist, who worked under the influence of the current belief in the constancy of species. But his enormous output of accurate description and of delineation of the most varied forms, has provided a sure basis upon which the more modern seeker after phyletic lines may proceed.

There have been few if any writers on botanical subjects so prolific as Sir William Hooker, and probably none have ever equalled him in the number and accuracy of the plates which illustrated his writings. Sir Joseph Hooker estimates the number of the latter at nearly 8000, of which about 1800 were from drawings executed by himself. The remainder were chiefly from the hand of Walter Fitch, who acted as botanical limner to Sir William for thirty years, showing in the work fidelity, artistic skill and extraordinary rapidity of execution. The numbers quoted give some idea of the magnitude of the results.

For the purpose of a rapid review of the chief writings of Sir William's later years, they may be classified under three heads, viz. (1) Journals, (2) Floristic works, and (3) Writings on the Filicales. Taking first the Journals, one of the most remarkable features about them is the apparent variety and number of the enterprises on which Sir William engaged: this is, however, explained when they are pieced together as they will be found below. His connection during 45 years with large and growing gardens, into which the most varied living specimens were being drafted in a constant stream, put him in possession of a vast mass of facts, detached, but needing to be recorded. The materials were thus present for that type of publication styled a *Botanical Miscellany*. The majority of the serials which he edited took this form, and though published under various titles, dictated in some measure by the source of their publication, more than one of them was a mere continuation of a predecessor under a different title. The first of them appeared under the name of the *Exotic Flora*, in three volumes (1823–7), with 232 coloured plates illustrating subjects from the Gardens of Glasgow, Edinburgh, and Liverpool. But owing to his taking up in 1827 the editorship of the *Botanical Magazine*, then in a critical position, the *Exotic Flora* ceased, and its materials swelled the pages of the more ancient serial, with which he was connected till his death.

To those not intimately acquainted with the other serials edited by Sir William, their relations are difficult to trace. But Sir Joseph Hooker has given their titles in series, with their dates, as follows:

> *Botanical Miscellany.* 3 vols. 1830–33.
> *Journal of Botany.* 1 vol. 1834.
> *Companion of the Botanical Magazine.* 2 vols. 1835–36.
> *Jardine's Annals of Natural History.* 4 vols. 1838–40.
> *The Journal of Botany* (*continued*). Vols. II.–IV. 1840–42.
> *The London Journal of Botany.* 7 vols. 1842–48.
> *The Companion of the Botanical Magazine.* (New Series. 1845–48.)

London Journal of Botany and Kew Gardens Miscellany.
9 vols. 1849–57.

From this list it appears that throughout a long term of years, though under varying titles, the stream of information gathered chiefly through garden management was edited and published, taking the form of 28 volumes, with 556 plates. The "Floristic" works of Sir William Hooker began with the second edition of Curtis's *Flora Londinensis*, in five folio volumes, upon which he worked from 1817 to 1828. He contributed a large proportion of the plates from his own drawings, while the descriptions throughout (excepting those of the plates on Algae and Fungi by R. K. Greville) were enlarged, and rewritten by him. He was in fact the real author of the work, which, however, was so badly edited—even the letter-press was not paged—that citation of it was impossible, and it never took its proper place as a scientific work. Sir Joseph Hooker points out that the second edition was not properly styled *Flora Londinensis*, since it included many species which are not indigenous anywhere near London. But these were the lapses of the editor, not of the author and artist. Minor works were the accounts of the plants collected on Parry's and Sabine's Arctic voyages (1823–28), but the *Flora Boreali Americana* was a more important undertaking. It appeared as two quarto volumes (1829–40), in which 2500 species were described with numerous illustrations. It was based on the collections of various travellers, and included ferns and their allies. In 1830 came the first edition of the *British Flora*, a work which was continued through eight editions, the last being in 1860, and it contained 1636 species. The botanical results of Beechey's voyage in the "Blossom" to the Behring Sea, the Pacific Ocean, and China were produced jointly with Dr Walker-Arnott in 1830–41, as a quarto volume, with descriptions of about 2700 species, and notable for the diversity of the floras included. In 1849 the *Niger Flora* appeared, dealing with the collections of Vogel on the Niger expedition of 1841. But the most remarkable of all these floristic works was the great series of the *Icones Plantarum*. It was initiated in 1837 for the illustration of *New and rare plants selected from the Author's Herbarium*,

and was continued by him till his death in 1865. Owing to
the munificence of Bentham's bequest to the Kew Herbarium for
its continuance and illustration, it remains still as the principal
channel for the description and delineation of new and rare
plants from the Kew Herbarium. The fact that the number of
the plates is now about 3000 gives some idea of the magnitude
of this work, which was started by Sir William Hooker in the
later days of his Glasgow professorship.

It might well be thought that the production of the works
already named would have sufficed to occupy a life-time, especially when it is remembered that they were produced in the
intervals of leisure after the performance of the official duties of
a professor, and later of the Director of the growing establishment at Kew. But there still remain to be mentioned that noble
series of publications on the Filicales, which gave Sir William
Hooker the position of the leading Pteridologist of his time.
The series on ferns began with the *Icones Filicum* (1828–31) in
two folio volumes, with 240 coloured plates by R. K. Greville,
the text being written by Hooker. The same authors again
cooperated in the *Enumeratio Filicum* (1832), a work projected
to give the synonymy, citation of authors, habitat, and description
of new and imperfectly known species. But it only extended to
the first 13 genera, including the Lycopodineae, Ophioglosseae,
Marattiaceae, and Osmundaceae, and was then dropped. Here
may be conveniently introduced a number of volumes, which
were for the illustration of ferns, but not systematically arranged.
They were issued from time to time, and collectively give a
large but not a coordinated body of fact. They were, the *First
Century of Ferns*, issued in 1854; the *Filices Exoticae* in 1859;
a *Second Century of Ferns* in 1861 ; *British Ferns* also in 1861,
and *Garden Ferns* in 1862.

There still remain to be mentioned three great systematic
works on ferns, each of which is complete in itself, viz. the *Genera
Filicum*, the *Species Filicum*, and the *Synopsis Filicum*. The
first of these was the *Genera Filicum* (1838–40), a volume issued
in parts, royal octavo, with 126 coloured plates illustrating
135 genera. It goes under the joint names of Francis Bauer
and Sir William Hooker, the latter being described on the

title-page as Director of Kew. But the preface is dated May 1, 1838, from Glasgow, and it was printed at the University Press. The title-page further states that the plates were from the drawings of F. Bauer, but Sir Joseph Hooker points out (*l.c.* p. cviii), that "of the whole 135 genera depicted I think that 78 are by Fitch." Sir William in the preface states that "The plates have all been executed in my own residence, and under my own eye, in zincography, by a young artist, Walter Fitch, with a delicacy and accuracy which I trust will not discredit the figures from which they were copied." The result is one of the most sumptuous volumes in illustration of a single family ever published. After 70 years it is still the natural companion of all Pteridologists. At its close is a synopsis of the genera of ferns, according to Presl's arrangement, which Sir William describes as "the most full and complete that has yet been published." But in the preface he remarks that Presl "has laid too much stress on the number and other circumstances connected with the bundles of vessels in the stipes, which in the Herbarium are difficult of investigation." This is a specially illuminating passage for us at a time when anatomical characters are becoming ever more important as phyletic indices. It shows that readiness of diagnosis was for him a more important factor than details of structural similarity.

In the preface to the *Genera Filicum* Sir William says, he "would not have it to be understood that the Genera here introduced are what I definitely recommend as, in every instance, worthy of being retained.... A more accurate examination of the several species of each Genus, which are now under review in the preparation of a *Species Filicum*, will enable me hereafter to form a more correct judgement on this head than it is now in my power to do." The five volumes of the *Species Filicum* thus promised, appeared at intervals from 1846 to 1864. The work is briefly characterised by Sir Joseph as consisting of "descriptions of the known Ferns, particularly of such as exist in the Author's Herbarium, or are with sufficient accuracy described in the works to which he has had access, accompanied by numerous Figures. This which will probably prove to be the most enduring monument to my father's labour as a systematist and descriptive pteridologist, is comprised in five 8vo volumes, embracing nearly 2500 species, with 304 plates by Fitch, illustrating 520 of these. It occupied much of the latter eighteen years of his life, the last part appearing in 1864." The work is a most extraordinary mine of detailed information. It is a condensed extract from his own unrivalled *Herbarium of Ferns*, with exact data of distribution, and collectors' numbers. Probably no family so extensive as this has ever been monographed by a single hand with such minuteness and exhaustive care. It is the classic book of reference in the systematic study of ferns. But as indicated in the preface to the Genera, the judgement as to which genera are "worthy of being retained" had been exercised. The result was the merging of a number of the genera of Presl, and others, into neighbouring genera. Though this was somewhat drastically done in the *Species Filicum*, it comes out more prominently in the work upon which he entered in the very last months of his life, viz. the *Synopsis Filicum*. This work was published in 1868 as an octavo volume, with 9 coloured plates, containing analyses of 75 genera. Sir Joseph tells us (*l.c.* p. 117) that "Upon this work my father was engaged up to a few days before his decease, and 48 pages of it in print were left on his desk, together with the preface and much matter in manuscript. After full consideration it appeared to me that, with the material in hand, the aid of the *Species Filicum* completed only three years earlier, and of the Fern Herbarium in perfect order, and named according to his views, a competent botanist should find no great difficulty in carrying on this work to its completion. Such a botanist I knew my friend Mr Baker to be, and also that he had made a study of Ferns, and accepted my father's limitations of their genera and species. I therefore requested that gentleman to undertake the work, which to my great satisfaction he has done. The *Synopsis Filicum* contains 75 genera, and about 2252 species, inclusive of Osmundaceae, Schizaeaceae, Marattiaceae, and Ophioglossaceae, which are not included in the *Species Filicum*." This work summarised the Pteridological results of Sir William Hooker's life. The total number of plates of ferns published by him is about 1210, embracing 1267 species, of which about 250 appeared under the joint authorship of Dr Greville and himself. These figures are in themselves sufficient evidence of the extent of his Pteridographic work.

It has been noted that the number of genera in the *Genera Filicum* was 135, maintained approximately according to the limitations of Presl in his *Tentamen Pteridographiae*: allowance has, however, to be made for 23 genera of Parkeriaceae, Schizaeaceae, Osmundaceae, Marattiaceae, Ophioglossaceae, and Lycopodiaceae, which were omitted in the *Tentamen*. But in the *Synopsis Filicum* there were only 75. It is true that the three genera of Lycopodiaceae were excluded also from the *Synopsis*, but still there is the wide discrepancy between 132 of Presl's genera as against 75 in Hooker's *Synopsis*. This at once indicates a salient feature of his method. He merged a large number of genera, ranking many of the smaller ones as subgenera under the more comprehensive headings. Doubtless the reasons for this were various. One was his mistrust of anatomical data, which it must be confessed Presl put too much in the fore-front. The very first sentence of the *Tentamen* runs thus "Vasa plantarum principale signum esse ex eo patet, quod exinde primaria divisio omnium plantarum exstitit." But occasionally Sir William explained his reason in a specific case. Thus in the question of Kunze's sub-genus *Plagiogyria* of the genus *Lomaria*, which Mettenius had raised to the dignity of a distinct genus, he explained his reasons for merging it into the genus *Lomaria*. Mettenius had laid stress upon various characters, but especially on the *oblique annulus* as distinctive. On this Hooker remarks "even should the capsules in all the species referred to *Plagiogyria* prove to be helicogyrate, yet the habit and sori are so entirely in accordance with true *Lomaria* that, unless the student has the opportunity of examining very perfect specimens, or unless he examines the structure of the annulus of the very minute capsules under the high power of the microscope, the genus cannot be identified. Kunze only proposed to form a group or section under the name of *Plagiogyria*, but even that would be found inconvenient to retain in a work whose main object is to assist the tyro in the verification of genera and species: and natural habit is often a safer guide than minute microscopic characters." Thus we see that in his method convenience of diagnosis is put before the use of important structural characters. I have recently found reason to uphold the opinion of Mettenius on this point, and to confirm *Plagiogyria* as a substantive genus.

Similarly, the genera *Lophosoria* and *Metaxya* will have to be detached from *Alsophila*: Prantl removed *Microlepia* from *Davallia* into his new family of the *Dennstaedtiinae*, where they are related with *Patania* (*Dennstaedtia*), which Hooker had merged into *Dicksonia*. Goebel also has detached *Hecistopteris* which Hooker had placed in *Gymnogramme*, and has placed it with the *Vittarieae*. These are all examples of the way in which further study is tending to reverse the excessive merging of genera, which Hooker carried out in the interest of diagnostic convenience.

The general conclusion which we draw from contemplating Sir William Hooker's work on the systematic treatment of ferns is that it was carried out consistently to the end under the influence of the current belief in the Constancy of Species. The methods were not phylogenetic, as they have since become under the influence of evolutionary belief. The problem seems to have been to depict and describe with the utmost accuracy the multitudinous representatives of the Filicales, and to arrange them so that with the least possible difficulty and loss of time any given specimen could be located and named. But the result is not to dispose them in any genetic order. Even the arrangement of the larger genera according to the complexity of branching of the leaves appears as a method of convenience rather than of genesis, and subsequent inquiry is tending to show that so far as such series really exist, they will require to be read in converse. Goebel, in his paper on *Hecistopteris*, remarks that "the systematic grouping of the Leptosporangiate Ferns, as it is at present, e.g. in the *Synopsis Filicum*, is artificial throughout; it is adequate for the diagnosis of Ferns, but it does not give any satisfactory conclusion as to the affinity of the several forms." He proceeds to say that "a thorough investigation, taking into account the general characters of form of both the generations, will be necessary before the naturally related groups, and their relations to one another, are recognised in the plexus of forms of the Polypodiaceae."

Such observations as these must not be understood in any sense of disparagement of the work of this great man. They are merely intended to indicate his historical position. The *Origin of Species* was, it is true, published some few years before the *Synopsis Filicum*. But we must remember that Sir William Hooker was already an old man. Few men over 70 years of age alter their opinions, and the labourer who had grown old under the belief in the Constancy of Species could not in a few brief years be expected to change the methods of thought of a long and active life. We must take Sir William Hooker as perhaps the greatest and the last of the systematists who worked under the belief in the Constancy of Species. Because we have adopted a newer point of view, and take into consideration facts and arguments which were never his, and come to different conclusions now, is no reason for valuing one whit the less the achievements of this great botanist.

His published work was just as much fundamental as was his official work. We have seen how he provided in Kew the means of indefinite development later, by constructing the coordinating machine with its collections and its libraries. In somewhat similar sense his publications were also fundamental. He did not himself construct. There is, I believe, no great modification of system or of view which is to be associated with his name. But in the wealth of trustworthy detail, recorded both pictorially and in verbal diagnoses, he has supplied the foundation for future workers to build upon, laid surely and firmly by accurate observation, and therefore durable for all time.

One remark I may make as to the effect of his work on the trend of botanical activity in this country. We have noted that anatomy was not Sir William Hooker's strong point. He and many of his contemporaries did not pursue microscopic detail, and indeed seem to have avoided it. He was, however, a dominating botanical influence of the middle Victorian period. May we not see in these facts, combined with the extraordinary success of the systematic work carried on by himself, or under his guidance, a probable cause of that paralysis of laboratory investigation which ruled in Britain till the early seventies? British botany was at that time almost purely descriptive. The revival came within 10 years of the death of Sir William, and it is well to remember that the immediate stimulus to that revival was given by a botanist, who became later the Director of Kew, and was allied by marriage with Sir William Hooker himself. I mean, Sir William Thiselton-Dyer. The stimulus had its result in the active development of anatomical and physiological study of plants, as we see it in this country to-day. For a time the swing of the pendulum in this direction was too extreme and exclusive. I remember very well an occasion when Sir Joseph Hooker said to me, " You young men do not know your plants." And it was true, though it may be added that few indeed, at any time, knew them in the full Hookerian sense. A saner position is gradually being attained. But even now the systematic study of Angiosperms receives far too little attention among us, and is an almost open field for the young investigator.

I would conclude with one word of advice, which naturally springs from contemplation of a life-work such as Sir William Hooker's. We sometimes see wide-reaching phyletic conclusions advanced by writers who we know have not specific knowledge of the groups in question. Let us learn from Sir William the importance of specific knowledge. It is only on such a foundation that sound phyletic argument can proceed. Let us always remember that it is better to carry out sound work on species, as he did, without theorising on their phyletic relations, than to promulgate phyletic theories without a sufficient specific knowledge of the families themselves. The former will probably be lasting work, the latter runs every chance of early refutation. Under the most favourable circumstances analytical work is as a rule more durable than synthetic. Sir William Hooker's contributions fall chiefly under the former head, and will be found to have a corresponding element of durability.

Plate XIII

JOHN STEVENS HENSLOW (1851)

JOHN STEVENS HENSLOW
1796—1861

By GEORGE HENSLOW

An all-round man—appointed Professor of Mineralogy at Cambridge in 1826, but succeeds Martyn in the Chair of Botany a year later— essentially an ecologist—his famous teaching methods—"practical work"—his wide interests—country life—the educational museum— village amenities.

THE scientific career and parochial life of the late Rev. Prof. J. S. Henslow, are described by my late uncle, the Rev. Leonard Jenyns, in his *Memoir*[1]. I propose adding and illustrating some of his more personal traits, habits and pursuits as a scientific man, and to deal especially with his educational methods. His studies in science were by no means confined to one branch, thus Geology was first ardently pursued in conjunction with Sedgwick. It was in a tour together in the Isle of Wight in 1819, that they proposed establishing a "Corresponding Society, for the purpose of introducing subjects of natural history to the Cambridge students." The outcome of this idea, which was subsequently abandoned, was the "Cambridge Philosophical Society," of which " Henslow, B.A. was elected secretary in 1821[2]."

Conchology and Entomology claimed his attention ; one of his first discoveries was the rare insect *Macroplea equiseti*, his identical "find" being figured in Curtis' *British Entomology*, while he found the bivalve *Cyclas Henslowiana*, so named by

[1] *Memoir of the Rev. John Stevens Henslow, M.A., F.L.S., F.G.S., F.C.P.S.* (J. Van Voorst, 1862).
[2] *Memoir*, pp. 17 ff.

Dr Leach, at Baitsbite on the Cam. His first and best collection of insects was presented to the Cambridge Philosophical Society. Other discoveries were made in after years, and are referred to by Jenyns.

On the death of Dr E. D. Clarke, he offered himself for the Professorship of Mineralogy. Chemistry, as well as the study of Minerals, now occupied his attention. He was only 26 years of age, and still B.A., when elected to that chair. At the age of 27 he published his *Syllabus of Mineralogy* in 1823, "A useful manual of reference to all persons studying Mineralogy, independently of the immediate circumstances which led to its publication[1]."

In 1827 Prof. Martyn died and Prof. Henslow was elected to the chair of Botany, being succeeded by Whewell on resigning the Professorship of Mineralogy. He now turned his attention to the study of Botany; but he never paid much heed to systematic botany, for his taste lay in the direction of what is now called Ecology. He then wrote "Botanists would rather receive one of our most common weeds from a newly-discovered or newly-explored country, than a new species of an already known genus. There are higher departments of Botany than mere collectors of specimens are aware of; for to ascertain the geographical distribution of a well-known species is a point of vastly superior interest to the mere acquisition of a rare specimen." *À propos* of this he made elaborate epitomes of the Botanical Geographies of De Candolle, and of the writings of Humboldt, Poiret and others. His MS. is not unlike a forerunner of Schimper's *Botanical Geography* of to-day. He thus expressed himself in the Introduction to his *Descriptive and Physiological Botany* (1836):—in the second section headed *Botany*..."This enquiry should extend as well to the investigation of the outward forms [of plant organs] and the conditions in which plants, whether recent or fossil, are met with, as to the examination of the various functions which they perform whilst in the living state and to the laws by which their distribution on the earth's surface is regulated." Again, in the Preface to the *Flora of Suffolk* by himself and E. S. Skepper,

[1] *Memoir*, p. 29.

he wrote:—"We had thought of saying something in regard to the Geographic distribution of the species, but found our material insufficient for treating this question to advantage." As an alternative he suggests interleaving the 'Catalogue,' as the book was also called, in which observers could add observations on the Geological formations and superficial soils upon which each species grows, e.g. Chalk, the Crags, Gravels of post-tertiary period, &c. as well as maritime, marshy, boggy, healthy and cultivated soils[1]."

Though he wrote against *mere* collecting, he was an insatiable collector himself; but it was always with some definite, useful and generally educational purpose, and the *best* of his collections invariably went to museums, especially those of the Philosophical Society of Cambridge, of Kew and of Ipswich. The first still has the fishes he collected at Weymouth in 1832, solely for his brother-in-law L. Jenyns, the author of *The British Vertebrate Animals.*

One of the first things to which his attention was directed was the Cambridge Botanic Garden. It was far too small and in the centre of the town, where the scientific buildings are now erected. He urged the necessity of a new one, but it was not till 1831 that the present site was secured; the first tree, however, was not planted until 1846.

His educational method of teaching was totally different from the mere instructional method of all previous lecturers. To cram up facts was the students' duty in the Medical schools, where botany was supposed to be taught. To learn by their own discovery was his new method, and so each student educated himself by examining and recording plant structures first seen by his own dissections. Having long been in the habit of observing himself, he was early convinced of the importance of practical work and he always had "demonstrations," as he called them, from living specimens. Each member of the class had a round wooden plate for dissecting upon. He had only sixteen lectures to give, but he succeeded in arousing an enthusiasm in

[1] Such are the "Conditions of Life," upon the "Direct Action," of which Darwin lays so much stress, as resulting in "Definite Variations...without the aid of selection." (*Var. of An. and Pl. under Dom.* II. p. 271 ff.; *Origin* etc. 6th ed. p. 106, etc.)

some, and interest in all who attended, and thus many came besides undergraduates, as Dr Ainslie, the Master of Pembroke. The value of "practical work" put a stop to cram, and he was the first to introduce the examination of flowers, not only at Cambridge but for the degrees in the University of London. "He insisted," wrote Dr Hooker, "that a knowledge of physiological botany, technical terms, minute anatomy, &c. were not subjects by which a candidate's real knowledge could be tested, for the longest memory must win the day, the less did it test the observing or reasoning faculties of the men. He, therefore, insisted in all his examinations that the men should dissect specimens, describe their organs systematically and be prepared to explain their relations, uses and significations in a physiological and classificatory point of view; and thus prove that they had used their eyes, hands and heads, as well as their books[1]."

His natural bent and interest were in the investigations of the phenomena of plant-life, e.g. the colours of flowers, the laws of phyllotaxis and what would now be called biometrical studies, e.g. of the variations in the leaves of *Paris* and the cotyledons of the sycamore, hybridization, teratology and the origin of varieties, etc. The geographical distribution of plants and the effects of external agencies upon them were also specially studied, as is recorded in the note-book mentioned. He was thus a genuine Ecologist without knowing it. He published about 50 papers on botanical subjects during his professorship from 1825 to 1861, in which he was more than once the pioneer of special branches of study since taken up, as in the above mentioned hybridization and varietal differences under cultivation, etc.; for experiments were made on the specific identity between the Primrose, Oxlip, Cowslip and Polyanthus. He raised many varieties, which were often permanent or "Mutations"; though sometimes reversions appeared, concluding that when one form thus changed to another that was sufficient proof of identity.

Though his occupations were necessarily much changed at

---

[1] Quoted in *Memoir*, p. 161.

Hitcham, of which he became the Rector in 1838, from those at Cambridge, he by no means neglected science; but he utilized it in different ways. Thus having a good knowledge of chemistry, he endeavoured to make the farmers interested in more scientific methods of farming than they had been accustomed to. He gave lectures on the fermentation of manures and he wrote fifteen "Letters to Farmers," first published in the *Bury Post* and then separately. He even proposed that they should make experiments themselves. For this purpose he issued schedules to about 70 farmers who asked for them.

The experiment was to test Liebig's suggestion that gypsum should be added to manure heaps to fix the ammonia. Unfortunately there is no record of the results[1].

The most important discovery from an industrial point of view, due to his knowledge of Geology, was undoubtedly that of the phosphate nodules known in the trade as "Coprolite," at Felixstowe in 1843, when he and his family were staying there. The cliffs are formed of "London clay," topped by the "Red Crag," between which is a bed of rolled, brown pebbles, once, with the crag, forming an ancient beach. Where the white "Coralline[2]" Crag occurs, the pebble bed lies *below* it. This accounts for the fact that it contains remains of Miocene animals, such as teeth of the Hipparion, or ancestor of the horse.

As the sea is always encroaching, the cliff has much "talus" in places, upon which was strewed the debris from the crag, including vast quantities of pebbles. Observing that they often contained a shark's tooth or other organic remains, he suspected that they might be composed partly of phosphate of lime. This proved to be the case, for the first analysis made by Mr Potter of Lambeth showed 54 % (1844). He communicated the fact to Mr, subsequently Sir, John Bennet Lawes, who desired a ton of nodules to be forwarded to him for experiment.

---

[1] On enquiring at Rothamstead, Mr Hall has kindly informed me that a "good deal of attention was given in Germany to this and other possible materials for the conservation of the nitrogen; but the general result was adverse to their employment."

[2] A misnomer, as the coralloid organisms are Bryozoa.

This led to their becoming a recognised article of trade. Large fortunes have been realised in Suffolk by owners of land containing the nodule bed, though frequently occurring at a considerable depth.

In 1848 he advocated the use of phosphate nodules in the "Greensand" beds of Cambridgeshire. These also soon became a commercial commodity.

In 1849, Professor Henslow delivered the inaugural address on the foundation of the Ipswich Museum, the object being, for "Giving Instruction to the working Classes in Ipswich in various branches of Science and more especially Natural History." It affords the best example of his views generally upon the uses of Science, not only as being of indisputable value in all useful arts, but as a means of education by dispelling the then prevailing ignorance and harmful prejudices rife in those days, even among men learned in other subjects at our Universities.

He illustrates his remarks from the chief sciences, as in Astronomy, by its importance in understanding the laws of storms and tides, which Whewell was then studying. Agriculture was touched upon, in showing the importance of a knowledge of Vegetable Physiology, and illustrated by the parasites, yellow Rattle and Wheat-rust. He insisted upon the educational value of accuracy, demanded of the scientist, and the avoiding *a priori* assumptions and hastily drawn deductions from insufficient data. But even the philosopher himself does not always escape from the imputation; for the farmers at Hitcham were firmly convinced that the "Piperage" or Barberry *itself* blighted the wheat. The Professor could not convince them that the red colour of the spots on the leaves of the bush was not due to the same fungus as that on the wheat. Indeed, he observes (in a MS.): "It is not likely (as some suppose) that it is due to the influence of *Æcidium berberidis*." We now know that the farmers were nearer the truth and the botanists were wrong. But one point the Professor established —and I possess his dried specimens to this day—and that was, that the "mildew," a black fungus, subsequently arises from the same substratum or mycelium as the rust. The mildew, then, throws off orange-coloured dust-like "spores," which attack the Barberry, and so the cycle is completed[1].

I still possess his dried specimens of other species of *Æcidium* attacking various kinds of plants, which he collected for comparison with that of the Barberry.

As abortive attempts to find coal had been made in some counties, he pointed out the value of Geology in at least intimating where coal was possible and also where it was impossible. It was not, he said, that a "little knowledge is a dangerous thing," as no one would become learned if he did not begin with a little, but it was the hasty deductions that were valueless and often dangerous.

As a practical illustration of this under the false assumption that the roots made the "bulb" of mangold-wurzel, he noticed the common practice of stripping off the leaves of plants, and explained to them that unless they were required for fodder, it was a wasteful practice, as the leaves (and not the roots, as they supposed) were the makers of the "bulbs." Indeed, in 1860, Prof. Jas. Buckman proved that it lessens the weight of mangold-wurzel by nearly one half.

Science was not even shut out at the Hitcham Horticultural Society's Exhibitions, for he always had his own marquee erected and a large board over the entrance with " The Marquee Museum " upon it, the letters being composed of Hitcham freshwater mussel shells. During the day of the show, he would deliver "lecturets" from time to time on the various specimens exhibited.

The following are samples of the latter. Cases of land and fresh-water shells of Hitcham. Photographs of microscopic objects enlarged, including the *first* ever made, by the Rev. H. Kingsley, Tutor of Sidney College, Camb. in 1855. A case containing living specimens of the smallest British Mammal, the harvest mouse. Pearls from British molluscs. The slow-worm

---

[1] In his printed *Report on the Diseases of Wheat*, written for private circulation only, he has added in MS.—"In specimens of true mildew, the *three forms—Uredo rubigo, U. lincaris* and *Puccinia graminis*, coexist simultaneously in the same sori, as well as numerous intermediate forms, which establish the specific identity of these fungi." *U. rubigo-vera* is now regarded as a form of *Puccinia rubigo-vera* and *Æcidium asperifolii.*

and viper in spirits, to show their differences. Hornets' and wasps' nests, naturally mounted, taken by himself, etc.

The Monday afternoon lessons in botany in the village school-room, held after school-hours, were always remarkable for the enthusiasm exhibited by the children. They were perfectly voluntary, but none was admitted to the Third Class until the child had learnt to spell correctly thirteen terms of classification of the classes, divisions and sections. On entering the class they at once began to fill up the "Floral Schedule[1]."

The botanical lesson included :—

1st—Inspection of specimens, anything special noticed and explained.

2nd—"Hard word" exercises. Two or three words (botanical terms) given to be correctly spelt on the next Monday.

3rd—Specimens examined and dissected and floral schedules, traced on slates, to be filled up. Marks allowed for accuracy, etc.

4th—Questions on the plant "organs."

Botanical excursions were made for those only who had received a sufficient number of marks.

The First Class came at certain times to the rectory on Sunday afternoons after Divine Service; when objects of natural history were shown and "such accounts given of them as may tend to improve our means of better appreciating the wisdom, power, and goodness of the Creator[2]."

A printed list of all the wild flowers in Hitcham was always suspended in the school-room, and a rack for named phials, which the children had to keep supplied with flowers as they came into blossom. Of course, little rewards were given to those who first found a flower and those who supplied the greater number, etc.

One of the exhibits of the Horticultural Shows was the collections of wild flowers made by the children. In addition, a public examination in botany was held, and a stranger would

---

[1] From the Professor's display of the methods he adopted of teaching Botany in schools, now in the South Kensington Museum, and Prof. D. Oliver's *Lessons*, etc. based on MS. left unfinished at my father's death, the floral schedule has been adopted in schools, not only all through the British Isles, but the Colonies as well.

[2] A completer account will be found in Jenyns' *Memoir*.

often find it a difficult matter to puzzle one of the best pupils, not merely as to the name—a trivial matter—but as to the structure of the flower itself.

The Government Inspector in 1858, wrote as follows in his Report :—" Extra subjects, pretty fair, and among them Botany, *excellent*; this last being most thoroughly yet simply taught, and by such a system that there can be no cram.   As far as a child goes, it must *know* what it does.   *The good moral effect of this study on the minds of the children is very apparent.*"

In those days, I am speaking of the "fifties," Darwin had not enlightened us as to the wonderful adaptations of flowers for fertilization by insects.   This adds enormously to the interest of the study—as the present writer soon found with village children of the parishes in which he has lived, and taught them botany—but even without that attraction the Hitcham children were intensely enthusiastic.

The Professor also taught them how to dry plants.   The village Herbarium, containing all the plants growing wild in Hitcham, was entirely màde by them.

It may be asked by cynics, "What can be the *use* of teaching science to such children ?"   It is not the mere fact that a child knows the structure of a rose, but it is the training in *accuracy of observation, mind and habit*, which the minute and close observation demands, i.e. if it be properly taught, and to secure that, is all important in children, who are naturally inattentive and inaccurate in consequence.   In teaching them botany as described above, the child is trained to avoid this bad habit in an interesting way, because inattention is solely due to want of interest.

The Ipswich Museum was a great source of pleasure to him. As President he carried out his plan of making it a "typical" museum, never letting it degenerate into a mere show, as so many country museums are, or at least used to be.

The Ipswich Museum has been a model for all others in that typical series of fossils, etc., are exhibited in the visible cases, all others being relegated to drawers, for students to examine. In allusion to the uses of Museums in his inaugural address referred to above, he remarked :—" Our collections should be viewed as the means of assisting us in the acquisition of real knowledge, and not merely to be gazed at as raree shows, or as only valuable in proportion to the number or scarcity of the objects they contain."

Of course, periodical lectures were delivered by the Professor at Ipswich, and he was a most lucid and admirable exponent.

He was the first to maintain that in museums of animals, they should, whenever possible, as, e.g. with birds, be represented in their natural conditions. With this object he collected nests with the boughs, or whatever it was in which they rested. Since then this plan has been admirably carried out at the Natural History Museum, South Kensington. He also supplied several museums with wasps' and hornets' nests with their surroundings. The plan he discovered most convenient for taking them, was to saturate tow with spirits of turpentine and place it at night in the hole, covered over with an inverted and corked flower-pot. The nest could then be dug up with impunity, as all the wasps were dead or torpid by the following morning. He always preserved the "pavement" or bottom-soil covered with stones which accumulated as the hollow for the nest increased in size. The nest was then suspended over it on rods to show the exact position. It was also half-dissected, to exhibit the interior, all the grubs having been carefully extracted. The village carpenter, the late Mr W. Baker, was a most enthusiastic assistant in taking and mounting the specimens.

When the potato famine occurred in Ireland in 1845—46, the disease was very prevalent in Hitcham. This induced the Professor to explain to his parishioners and others—for he published his recommendations—how they could utilise their rotten potatoes by extracting the valuable starch, which still remained sound within the tubers, even when these were refused by pigs. The process is so simple that it may be mentioned here. The potatoes must be *grated* (a piece of tin with holes punched through it will do); the pulp is then stirred with a stream of cold water through a hair-sieve. The *brown* water must be allowed a few minutes for the starch, carried through, to settle. The water is poured off, and the layer of starch must be stirred up and washed with fresh cold water. This may be done two or three times, till it becomes perfectly white. It must then be carefully dried in the sun or in a warm room (our method was to hang it up in small muslin bags in the kitchen); the bags must be repeatedly "kneaded" to prevent its clotting. When perfectly dry, it will keep for any length of time. Of course, it is precisely the same thing as sago, tapioca, cornflour, arrowroot, etc. and can be used like them. All our potatoes in the Rectory garden were rotten, but we recovered at least two sacks of starch. I remember taking a large sponge-cake to school, more or less made with this potato-flour, and making my reverend master somewhat incredulous by telling him it was made out of rotten potatoes!

Professor Henslow printed and circulated the receipt for the extraction of starch, in the village; so that several, who thought it worth while, obtained considerable quantities of starch.

In one of his lectures, dealing with this subject, he pointed out how a good basin of "arrowroot" can be made in ten minutes from two or three fair-sized potatoes; for as soon as the starch has been thoroughly "washed," it is ready for the boiling milk. It is essential the milk or water should be actually boiling, or the granules of starch do not burst and so make the required "jelly."

The school children of Hitcham were by no means left out in the cold as to the knowledge of natural phenomena. They were early instructed as to the harmless nature of toads and slow-worms, which were very abundant, on the one hand; and of the danger of handling a viper, on the other. This last is the only poisonous reptile in England, and easily recognisable by the lozenge-shaped marks down the back. Having specimens in spirit, they had no excuse for confounding them; but, as always happens with children, if there is an alternative of any sort between which they are well taught the difference, some one is sure to get them transposed in his memory. Consequently, a boy came up to the Rectory with his arm greatly swollen; he had been bitten by a viper which he had taken up, thinking it was a slow-worm, *because*, as he said, it had the marks along its back!

Besides the tiny harvest mice, he at one time possessed for some two or three years two "pet" Jersey toads, or the great *crapaud*. They were kept in a wire-gauze cage, and it was our delight as children to feed these monsters every morning. A butterfly net swept over the lawn was sure to secure all sorts of flying and jumping creatures. The lid of the cage being lifted up, the net was turned inside out over the toads, and quickly closed. Then began the matutinal breakfast. They would never notice anything that did not move. Seeing, however, say a grasshopper, stir, the toad would stalk it like a cat after a bird; and when within tongue-shot, out came its long tongue like a flash of lightning, and the grasshopper vanished in the flash. Worms were a great delight. Snapping up one in the middle, the two ends were carefully cleaned from earth by passing them between the toes two or three times; then followed a mighty gulp, and all was over.

Shell-traps were always laid about the grass, consisting of slates, under which there would generally be found a various crop of sorts. I have now two glass cases containing all the shells, land and fresh-water, of Hitcham, mounted by the Professor himself. A reward was offered for every specimen of a Hélix with the shell reversed. They are very rare, but one was brought by a little boy who discovered it, for he found he was unable to get his thumb into the opening the right way when playing at "conquerors." So he got the only sixpence earned in twenty-three years that the Professor was incumbent of Hitcham. The collection of butterflies was always being added to; now and then a rare one would appear at Hitcham, as, e.g. the Camberwell Beauty. The Professor was walking in the Rectory garden with the late Judge Eagle, of Bury St Edmunds, when one settled on a wall. Mr Eagle stood sentry while the Professor ran indoors for his net. It need hardly be added that the specimen still rests in the collection, which passed into the possession of his son-in-law, the late Sir J. D. Hooker, F.R.S., etc.

I cannot do better than conclude with my uncle's words at the end of his *Memoir*:—"When a good man dies the world does not cease to benefit from those labours of love which he undertook for his fellow men. Though personally removed from them his example remains; his voice too, is still heard in the lessons left to be handed down to those who come after him. The influences of Professor Henslow's teaching have been felt in other places than those in which he himself taught, they have borne fruit far beyond the obscure neighbourhood in which he first sowed the good seed, and who shall say to what further results they may not grow in years to come, bringing honour to his memory, and what is far more, glory to God? '*A word spoken in due season, how good is it!*'"

JOHN LINDLEY

1799—1865

By FREDERICK KEEBLE

Rise of Systematic Botany—Lindley's place—early history—services to Horticulture—Professor at University College, London—*The Gardeners' Chronicle*—*Theory and Practice of Horticulture*—*The Vegetable Kingdom*—Orchids—his interest in Fossil Botany—personal characteristics.

Introduction.

THE first half of the 19th century is a brilliant epoch in the history of botanical discovery. During that period the foundations of plant-anatomy were laid afresh with the cell as the builders' material. The discovery of sarcode or protoplasm electrified the scientific world and excited the attention of the philosophical novelist—as readers of *Middlemarch* may remember. The nucleus, the only and true *deus ex machina* of many a modern botanist, was recognised as an organ of the cell.

Biochemistry came into being and, with Liebig as foster-parent, grew into modern Physiology. The natural system of classification proclaimed by Jussieu put to rout the old established Linnean system and the enunciation of the theory of Natural Selection brought the epoch to a dramatic close.

In the constructive work of this period British botanists played a distinguished part, and it was due preeminently to them that the transition from the old artificial system to the new natural system took place so speedily and completely.

The group of men to whose labours this great change was due include Hooker, Brown, Bentham and the subject of this sketch, John Lindley. Nor from this brief list may the name

Plate XIV

JOHN LINDLEY (1848)

of Sir Joseph Banks, "the greatest Englishman of his time," be omitted. The commanding position to which these men attained in the world of science was of course due, primarily, to their ability and—equally of course—to circumstance. The great wars were over and in the peaceful years men were free to turn their energy to constructive purposes. Horticulture— ever a British art—became unreservedly popular. Explorers and collectors, encouraged and assisted by Banks and others, sent home rich supplies of new or rare plants and thus provided British systematists with a vast array of material for their work of reconstructing the flora of the world. Such brilliant use was made of opportunity that our country took the lead in systematic botany.

The activity of the collector, the generosity of the patron and the labour of the systematist led not only to a general advance in methods of classification but also to a very special advance in the knowledge of what is, in many ways, the most interesting group of plants on the face of the earth—the Orchidaceae. Among the plant-treasure from India, Australia and Malaya were large numbers of epiphytic orchids. The problem of cultivating such strange and fascinating plants challenged the skill of the gardener. The "fancying" instinct, latent in every Englishman and curiously characteristic of the race, was evoked by the bizarre form of these plants. Orchid-growing became the hobby of the well-to-do. Gardeners with no knowledge of science and regardless of text-book dicta on sterility, proceeded to raise the most marvellous series of hybrids— bi-generic, tri-generic, multi-generic—which any sane and scholastic botanist would have declared to be impossible.

Brown, Blume and above all Lindley threw themselves with enthusiasm into the task of discovering the clues to the classification of these plants, the form of whose flowers transgress so glaringly the rules of morphology—dimly surmising perhaps that if the key to evolution is ever to be found it will be discovered by the study of the group of plants which appear to represent evolution's latest prank.

In building up the new system of classification of the

vegetable kingdom in general and of orchids in particular, Lindley bore a conspicuous part; and were these his only contributions to the advancement of botanical science, his biographer might find the task of writing his life one of no very great difficulty. When however he discovers the many other varied aspects of Lindley's activities, the biographer may well despair of presenting a fair picture of the scientific life of this remarkable man. Professor of Botany in University College, London, "Præfectus Horti" to the Society of Apothecaries, officially attached to the Royal Horticultural Society and responsible for the management of its gardens, and in no small measure for its very existence, Lindley yet found time to become easily the greatest scientific journalist of his age. For nearly 25 years he edited the *Gardeners' Chronicle* and did more than any other man to keep the science and practice of horticulture on good terms with one another. To those of us who know how generally the cares of organisation give excuse for slackness in research, Lindley's indomitable activity, both in administration and in investigation, becomes indeed impressive and inspiring. Lecturing, drawing and describing new genera and species, revising the vegetable kingdom, writing memoirs, text-books, articles, directing the gardens at Chiswick, fighting officialdom and obstruction, building up a great herbarium and discharging a dozen other duties would seem to have made up the daily life of this man of amazing vigour. Till he was 50 years of age Lindley never knew what it was to feel fatigue; at 52 he took his first holiday; but the continuous strain of half a century had exhausted him beyond recuperation. He rallied, set to work again, again broke down and died at the age of 67.

To sketch in rapid outline and to admire to the full, John Lindley's life is not difficult even to the modern botanist whose life is passed in the cloistered calm of the laboratory; but to give a discriminating account of the chief of Lindley's services to science is well-nigh impossible for any one man: certainly I could not have undertaken it unaided. Good fortune and friends however rendered the attempt unnecessary. In the first place, Lord Lindley, when he knew of this project, put at my disposal in the kindest manner possible an outline of John

Lindley's career which he had written under the title of "Sketch of my Father's Life: written for my sons, daughters and grandchildren." In what follows I have made free use of Lord Lindley's manuscript. In the second place, Mr W. Botting Hemsley has had the great kindness not only to supply me with much valuable information of which he was possessed concerning Lindley's scientific work but to examine manuscripts, letters, etc. at Kew bearing thereon and to allow me to make use of the results of his interesting investigations.

Hence my task has become merely that of an editor whose chief duty is to fit the material provided by two distinguished contributors into the prescribed space. Whatever credit is due to this first attempt to sketch the career of Lindley, belongs to these two gentlemen whose remarkable kindness I have great pleasure in acknowledging.

Outline of Career.

John Lindley was born on February 5, 1799, in Catton near Norwich. His father, George Lindley, who came of an old Yorkshire family, conducted a large nursery and fruit business in Catton. To the facts that John Lindley became in early years an accomplished field botanist and also learned much of practical horticulture may be ascribed the close touch which he maintained throughout his botanical career with the practical side of botany. It is not too much to say that John Lindley was the unique representative of a class of man which he himself declared had never existed, namely one which combined the qualities of a good physiologist with those of a practical gardener of the greatest experience. John Lindley's youthful ambition was however to be not a savant but a soldier, and though, owing to the inability of his father to buy him a commission, that ambition was not fulfilled, the instinct which prompted it found frequent expression throughout Lindley's life. As his career demonstrates, he was a first class fighting man. The curious may find in the pages of the *Gardeners' Chronicle* records of the combats which he waged on behalf of horticulture and we shall have occasion presently to refer to the most important of all his campaigns in the cause of science.

When John Lindley was about 19 or 20 years of age his father's affairs became involved, and the son with an impulsiveness as just as it was foolish insisted, against the advice of friends, on becoming surety for the father. The mill-stone of financial anxiety thus early hung about his neck caused him trouble throughout his life.

Possessed of nothing but youth, a sound education, great natural ability and one good friend, John Lindley at the age of 20 left Norfolk for London. Thanks to a letter of introduction from the friend (Sir William Hooker) he obtained a post as assistant-librarian to Sir Joseph Banks. He thus gained access to a good library and became acquainted with a large number of men, both English and foreign, interested in scientific subjects. That he made the most of his opportunities is evident, for we find him at 21 a Fellow of the Linnean Society and a member of the Bonn Academy of Natural History. In 1822 began Lindley's long connection with the Horticultural Society, which he served first as Garden Assistant-Secretary, then (1826—1860) as Assistant-Secretary and finally as Secretary.

The portrait which accompanies this sketch is a reproduction of that painted by Mr Eddis, R.A., at the instance of friends of Lindley about the time of his resignation of the Secretaryship of the Horticultural Society.

The most conspicuous direct services rendered by Lindley to the Society were the laying out of the Society's garden at Chiswick and the organisation, with Bentham, of the celebrated flower-shows which have served as models for the exhibits of horticultural societies all over the world. Those who know how extraordinarily valuable, not only to horticulturists but also to botanists, are the periodical "shows" held by the Royal Horticultural Society, will be grateful to Lindley for the perspicuity which led him to replace the old and gaudy "fêtes" by these admirable exhibitions.

Lindley's Professorship of Botany in University College, London, dates from 1828 and was held for over a quarter of a century. Among those who attended his lectures were

Carpenter, Edwin Lankester, Griffith, Daubeny and Williamson. His lectureship to the Society of Apothecaries began in
1835, and in 1841 in which year the *Gardeners' Chronicle* was
founded, he became editor of that periodical. This post he held
till his death in 1865.

It might be supposed that the multifariousness and onerous-
ness of Lindley's official and routine duties left little time for
other work. Yet Lindley made time not only for scientific
investigation and for the writing of numerous monographs and
text-books; but also for a large and varied amount of public
work. In the Lindley correspondence preserved at Kew are
to be found letters and papers (official correspondence 1832—
1854) criticising trenchantly the mismanagement of the Royal
forests and recommendations on the selection and cultivation
of trees for the charcoal employed in the manufacture of gun-
powder.

Lindley, together with Hooker, acted as adviser to the
Commissioners of the Admiralty with respect to the planting
of the Island of Ascension.

The potato famine was the occasion of an official visit to
Ireland and led to a report by Lindley, Sir Robert Kave and
Sir Lyon Playfair which was the immediate cause of the Repeal
of the Corn Laws. As Sir Robert Peel told Lindley "in the
face of the Report, the repeal could no longer be avoided."
Thus the potato takes rank with the chance word, the common
soldier, the girl at the door of an inn that have changed or
almost changed the fate of nations.

Lindley and Kew.

But of all Lindley's public works that which he undertook
for the saving of Kew from destruction is of the most immediate
interest to botanists. In 1838 a small committee consisting of
Lindley, Paxton and J. Wilson (gardener to the Earl of Surrey)
were commissioned to report on the state of the Royal Gardens.
After exposing the incompetence and extravagance of the then
administration Lindley recommended that the Royal Gardens,
Kew, should be made over to the nation and should become the

headquarters of botanical science for England, its Colonies and Dependencies. Is it due to our lack of gratitude or to our mistrust of sculptors, that no statue of Lindley stands in the grounds of Kew? In 1840 John Lindley was able to write to Sir William Hooker: " It is rumoured that you are appointed to Kew. If so I shall have still more reason to rejoice at the determination I took to oppose the barbarous Treasury scheme of destroying the place; for I of course was aware that the stand I made and the opposition I created would destroy all possibility of my receiving any appointment." Having regard to the part which Lindley played in preserving Kew from the devastating clutches of the politicians it is but fit that that Institution should contain the most valuable of Lindley's scientific possessions, his orchid herbarium,—that his general herbarium is at Cambridge may be news to such Cambridge botanists as in the days of a decade or two ago learned Botany without such adventitious aids.

In 1864 Lindley wrote to the late Sir Joseph Hooker to say that he had made up his mind to sell his herbarium and would prefer that the Orchids went to Kew. There it is preserved, a monument of Lindley's skill and industry and of inestimable value to the systematist. Besides the actual specimens it contains coloured drawings of the flowers of all the species that came under his observation in the living state. In addition to the herbarium, Kew possesses a large amount of Lindley's scientific correspondence; letters to W. J. Hooker, 1828—1859 (230), 182 letters to Bentham and 35 to Henslow, and others to which reference has been made already: altogether an invaluable mass of correspondence, selections from which it is to be hoped may some day see the light of publication.

Lindley's skill with brush and pencil may be admired in the many plates which he executed in illustration of his various monographs. His skill with the pen deserves at least remark. Inasmuch however as nearly all the more distinguished of the old school of botanists, Hales, Hooker, Gray, to mention but a few, have in this respect a marked superiority over their successors, it is not necessary to labour the question of literary grace for either the moderns are indifferent on the subject or they may find on every hand models ready for their use. Two citations from the introductory pages of Lindley's classic, *The Theory and Practice of Horticulture*, must suffice to exemplify his incisive style—Le style c'est l'homme, and Lindley the man hated circumlocution and had no time to waste—"there are, doubtless, many men of cultivated or idle minds who think waiting upon Providence much better than any attempt to improve their condition by the exertion of their reasoning faculties. For such persons books are not written"; and again, with reference to the divorce in current literature between theory and practice, "Horticulture is by these means rendered a very complicated subject, so that none but practical gardeners can hope to pursue it successfully; and like all empirical things, it is degraded into a code of peremptory precepts."

Publications. "The Theory and Practice of Horticulture."

Though many aspects of Lindley's work must perforce be treated of in briefest form no sketch could have the slenderest value which did not take into account his chief works, *The Theory and Practice of Horticulture, The Vegetable Kingdom*, and the *Botanical Register*; nor from a survey no matter how brief may reference to his contributions to our knowledge of orchids be omitted.

The value of Lindley's great work on *The Theory and Practice of Horticulture* may be best gauged by the fact that as a statement of horticultural principles it is the best book extant. Though the botanist of the present day finds on perusing this work that physiological knowledge in 1840 was in a singularly crude state, and may rejoice at the rapid progress of discovery since the time when Lindley's book was written, yet the fact remains that few, if any, men at the present day could make a better statement of the physiological principles underlying practical horticulture than that presented by John Lindley.

Indeed it is a strange fact, and one worthy of the attention of our physiologists, that the gardeners are still endeavouring to puzzle out for themselves the reasons for their practices unaided by the physiologists. An interesting illustration of

this assertion may be found in recent issues of the *Gardeners'
Chronicle* containing correspondence from many of the leading
growers on the principles underlying the cultivation of the vine.
No physiological Philip has come as yet to their assistance!
Lindley's book had at once a great vogue on the Continent and
was translated into most European languages—Russian included; but it was not till its title was changed from *The
Theory... to The Theory and Practice... of Horticulture* that his
incorrigible fellow-countrymen, as shy of theory as a fox-glove
is of chalk, consented to buy it to any considerable extent.

It was doubtless due not only to Lindley's general services
to horticulture but also to the special service which he rendered
to that science by the publication of this work that led Lord
Wrottesley, President of the Royal Society, to say, when presenting Lindley with the Royal Medal, that "he had raised
horticulture from the condition of an empirical art to that of a
developed science."

"The Vegetable Kingdom" and "The Botanical Register."

That John Lindley was a man of fine judgment is indicated
by his own verdict that, except for *The Vegetable Kingdom, The
Theory and Practice of Horticulture* was his best book. That
verdict is sustained by posterity, as Mr Botting Hemsley de-
clares of the former work,—"This grand book must be classed
as Lindley's masterpiece. No similar English work was in
existence in 1846 when the first edition appeared, nor was there
in any language so encyclopaedic a work. Even now it is a
valuable book in a small botanical library as it is a mine of
information on points that are unchangeable. The work, as
set forth in the preface, originated in a desire on the part of the
author to make his countrymen acquainted with the progress
of Systematical Botany abroad during the previous quarter of
a century." Both in his books and in his lectures he adopted
the natural system of classification and did much to popularise
it though, as previously stated, his contemporaries Robert
Brown, the Hookers, and G. Bentham were equally powerful
adherents of the new system. To quote the picturesque if

somewhat immoderate language of Reichenbach "for a long time the youthful interloper found no favour on account of his having introduced in conjunction Scot Brown, Gray and the still youthful Hooker the natural system of the hated Frenchman; where the more numerous disciples of Linnæus had thought to pass their lives in the glory of pondering and admiring the great Swede." That Lindley was an early convert to this innovation is also proved by the fact that his inaugural lecture at University College startled many by its frank and thorough expression of the superficial character of the artificial system of classifying plants.

The third and last edition of *The Vegetable Kingdom* consists of about 1000 pages in small type with upwards of 500 illustrations. It contains an historical review of the various "Natural Systems" which had been prepared, beginning with John Ray's (1703) and ending with his own, which is used in the work. In this system Lindley divided plants into seven classes :— Thallogens, Acrogens, Rhizogens, Endogens, Dictyogens, Gymnogens and Endogens, and each class was subdivided into alliances or groups of Natural Orders to which he gave names of uniform termination, as Algales, Filicales, Glumales, Malvales, etc. This classification, though ingenious, is defective, as the author himself recognised. Though never adopted by other writers this fact did not prevent Bentham and Hooker from citing Lindley's work frequently in their *Genera Plantarum*. As Mr Botting Hemsley observes, Lindley, who in all questions of classification was both cautious and modest, seems to have been an evolutionist without knowing it. Thus in the course of discussion on the permanency of species he observes that " all the groups into which plants are thrown are in one sense artificial, in as much as nature recognises no such groups. As the Classes, Natural Orders and Genera of botanists have no real existence in Nature, it follows that they have no fixed limits and consequently it is impossible to define them....An arrangement then which shall be so absolutely correct an expression of the plan of nature as to justify its being called *the* Natural System is a chimera."

Owing to the fact that Hooker wrote the admirable and

favourable review of the *Origin of Species* which appeared in the *Gardeners' Chronicle*, it has been inferred that Lindley himself was not very well disposed toward the new theories; but Lord Lindley states that his father was much impressed by the *Origin*, said it would revolutionise botanical studies but that there were difficulties which would require elucidation before Darwin's theory could be regarded as completely satisfactory—surely a perspicacious judgment.

To turn to the woodcuts of *The Vegetable Kingdom* affords both pleasure and relief—pleasure on account of their excellence, relief to escape from the monotonous prettiness of modern process work.

Though space will not allow reference to other text-books and to innumerable minor publications—many of which may be found in the Lindley Library in the Royal Horticultural Society's headquarters at Vincent Square—a brief mention must be made of the *Botanical Register*. This periodical was founded in 1815, and so early as 1823 Lindley became a contributor to it; but it was not till 1829 that his name appeared on the title-page. From that time he was sole editor till 1847, when the *Botanical Register* ceased to appear; unable doubtless to stand against the *Botanical Magazine* which under the editorship of Hooker had passed from a moribund state into one of remarkable vigour which now, 125 years after its foundation, it still enjoys.

Orchids.

The magnitude of Lindley's work among his favourite group of plants, the Orchidaceae, deserves recognition by the general botanist. Botanical knowledge with respect to the group was in a very rudimentary stage when Lindley took up its study. Robert Brown and Blume were already engaged upon the investigation of orchids, but they relied mainly on herbarium material. Lindley, on the other hand, began with living plants and ended with living plants, though, as his herbarium testifies, he did not neglect dried specimens. A circumstance that favoured Lindley in these studies was the fact that William

Cattley, an early patron of Lindley, was one of the most successful of the early cultivators of epiphytic orchids.

The chief of Lindley's published contributions to the knowledge of orchids, apart from scattered figures and descriptions in the *Botanical Register*, the *Gardeners' Chronicle*, Lindley and Paxton's *Flower Garden*, the *Journal of the Linnean Society*, and in other serials and periodicals, are to be found in *The Genera and Species of Orchidaceous Plants*, 1830—1840, in which are described all the species (1980) known of 299 genera; *Sertum Orchidaceum* (1838); *Folia Orchidacea*, 1852—1855; and *The Vegetable Kingdom*.

It is unfortunate that no attempt has as yet been made to catalogue the species described by Lindley; but with regard to genera an approximate list of those proposed by him may be attempted, and is interesting as giving some idea of the extent and value of Lindley's investigations in the group.

In the third edition of *The Vegetable Kingdom* he estimates the number of orchid genera at 469. Bentham and Hooker (*Genera Plantarum*, 1883) admit 334, and new genera proposed since that date amount to 125. Pfitzer (Engler and Prantl, *Natürlichen Pflanzen-familien*, 1889) describes 410.

The following is a list of Lindley's genera, admitted by Bentham and Hooker, in the sequence in which they appear in the *Genera Plantarum*:

Physosiphon	Trichosma	Sophronitis	Acincta
Brachionidium	Coelogyne	Galeandra	Mormodes
Oberonia	Otochilus	Ansellia	Cycnoches
Oreorchis	Pholidota	Cremastra	Stenia
Sunipia	Lanium	Bromheadia	Clowesia
Cirrhopetalum	Diothonea	Govenia	Scuticaria
Megaclinium	Hormidium	Grobya	Camaridium
Trias	Hexisia	Cheiradenia	Dichaea
Drymoda	Pleuranthium	Aganisia	Trichopilia
Monomeria	Diacrium	Acacallis	Aspasia
Panisea	Ponera	Eriopsis	Cochlioda
Acrochaene	Pinelia	Warrea	Dignathe
Coelia	Hartwegia	Batemannia	Miltonia
Eria	Cattleya	Bifrenaria	Solenidium
Phreatia	Laeliopsis	Xylobium	Erycina
Chysis	Tetramicra	Lacaena	Abola
Anthogonium	Laelia	Lycaste	Trizeuxis
Earina	Schomburgkia	Chondrorhyncha	Ada

Sutrina	Diplocentrum	Gomphicis	Bicornella
Trigonidium	Cryptopus	Baskervilla	Hemipilia
Quekettia	Oeonia	Pelexia	Glossula
Zygostates	Mystacidium	Herpysma	Pachites
Phymatidium	Cirrhaea	Zeuxine	Herschelia
Centropetalum	Notylia	Haemaria	Monadenia
Doritis	Sertifera	Hylophila	Schizodium
Aëranthes	Tropidia	Drakaea	Forficaria
Uncifera	Pterichis	Burnettia	Brachycorythis
Acampe	Prescottia	Chloraea	
Sarcanthus	Pseudocentrum	Stenoglottis	

When it is remembered that Bentham, who elaborated the orchids for the *Genera Plantarum*, held broader views of generic limits than the majority of botanists, the fact that 114 or more than a third of the genera retained are Lindleyan is a striking testimony to the accuracy and range of Lindley's work in the group. Pfitzer in the work already cited retains 127 of Lindley's genera. In no other great family probably has one man left so large a mark as Lindley has left in the Orchidaceae. In this connection it may be added that 40 of Robert Brown's Orchid Genera and 50 of Blume's are retained by Bentham and Hooker.

The number of species of orchids known in his time Lindley doubtingly estimated at 3000. Collectors since that time have increased that number probably to 6000. The fact that about 1100 species of orchids are known from British India, outnumbering those of any other family by about 300, will doubtless surprise the majority of botanists.

Before closing this notice of a remarkable and versatile man some reference must be made to his pioneer work in the field of palaeobotany—a subject that has markedly advanced in recent times at the hands of Lindley's fellow-countrymen. In co-operation with Hutton there were published (1831—1837) the three volumes of Lindley and Hutton's *Fossil Flora of Great Britain*, an authoritative work, profusely illustrated with figures of the known fossils, and by no means entirely superseded at the present day. The introductory chapters to the volumes bear the mark of Lindley's handiwork, and that to volume III. contains the results of an extensive series of experiments carried out by Lindley to determine the capacity of various plants to resist the agencies of disintegration. These results have become

classic and are often referred to by subsequent writers on palaeobotany.

During the progress of the *Fossil Flora* Lindley amassed a
considerable collection of specimens, some of which have recently
come to light in the cellars of University College. He was
obliged however to abandon this branch of study as it threatened
to distract his attention from other departments of botany.

Personal Characteristics.

In as much as it is our custom to erect none but the slightest
and most casual memorials to our distinguished men of science
or of letters, there is reason to rejoice that the name of Lindley
is not inadequately commemorated.

The Lindley Library purchased in his honour and now
permanently attached to the Royal Horticultural Society bids
fair under the enlightened policy of that flourishing institution
to grow into a great collection of horticultural works. The
genus *Lindleya* is reminiscent to systematists of their great
colleague and the name of Lindley is known and honoured by
all our horticulturists. Of the man himself just so much may
be said as to give form to the mind's image of him.

He was of middle height, active, upright, with shoulders
somewhat sloping and of heavy tread. The sightlessness of one
eye gave to his resolute face a somewhat strange look. Simple
in habits, strenuous in work and perspicacious in judgment, John
Lindley was a warm hearted and generous friend, particularly
to young botanists. He was a powerful foe: altogether a
masterful and remarkable man. Not suffering fools gladly yet
with a humorous turn of mind: "I am a dandy in my herbarium,"
he once exclaimed to Reichenbach. Knowing no fear he could
not hope for much favour, and yet carrying his heavy load of
financial responsibility, he nevertheless won through to a wide
measure of contemporary recognition and an assured place in
the history of botanical science. To conclude with Reichenbach's
fine tribute "we cannot tell how long Botany, how long science,
will be pursued; but we may affirm that so long as a knowledge
of plants is considered necessary, so long will Lindley's name be
remembered with gratitude."

WILLIAM GRIFFITH

1810—1845

By W. H. LANG

Early training—medical appointment under the East India Company—his travels—the magnitude of his collections—his method of work—results of researches mainly published posthumously—the ovule and fertilisation—Santalum—Loranthaceae—Balanophora—Avicennia—his gymnosperm work illustrated by Cycas—discovery of the pollen-chamber —Rhizocarps and Liverworts—pre-Hofmeisterian work—Griffith's relation to his times.

IT might have been assumed that all the names of British botanists whose work has been or is to be considered in this course of lectures would have been familiar to their successors of to-day, even if their works were too often neglected for the last words of scientific progress in a summary of literature. The question has however been put to me by more than one botanist in the last month or two, "*But who was Griffith?*" That this should be possible seems in itself ample justification for including his name in this list of British botanists.

For Griffith has claims to be regarded as a great botanist. It is true that he failed to break through the limitations of his time and period—that he left no new and more correct general views to modify the science. But this is true of all his contemporaries, indeed it is true of most botanists. To recreate the department of a science in which a man labours requires a combination of ability and fortunate chance that is given to few.

Griffith had the ability, the power of independent observation, the readiness to speculate, the careless prodigality of labour. He did not however, in the fraction of an ordinary working life that

Plate XV

WILLIAM GRIFFITH (1843)

fate allowed him, attain that insight into more correct comparison of the plants whose morphology he studied which would have acted quickly on the mass of first hand observation he possessed.

It is well to be clear at the outset that it is the personality of William Griffith, his important detailed contributions to botany, and his achievement as a great working morphologist of his time that will interest us to-day—rather than his general views or any influence of these on the progress of botany. Griffith had the advantage or disadvantage of botany being his private study and not his profession. The motive force of his career was however his love of scientific work for its own sake.

William Griffith was a London botanist. He was the son of a London merchant, born on March 4, 1810, at Ham Common. Having finished school he began to prepare for the medical profession and was apprenticed to a surgeon in the West end of London. About 1829 he commenced attendance at the classes in the newly established University College. He had earlier in life shown an interest in natural history but was now specially devoted to botany. He attended Lindley's lectures, and also studied medical botany under Mr Anderson at the Apothecaries' Garden in Chelsea. There he obtained the Linnean Gold Medal given by the Society of Apothecaries. At this time also he was a frequent visitor to Kew Gardens where he was on good terms with the head gardener and also came under the influence of Mr Bauer the great botanical draughtsman of his day. Griffith was never tired of expressing his admiration for Bauer as an accurate observer. During his vacations Griffith made botanical excursions in England, carrying his light baggage and his equipment for collecting plants.

That the training that Griffith received in botany in the London University of that date was a sound one is shown by his power of facing the most various problems when cast on his own resources immediately at the close of his University training. The soundness of his training is further shown by the small pieces of original work he had published before leaving England at the age of 22. Not only had he made some of the illustrations for Lindley's *Introduction to Botany* and had described the flower and the structure of the wood of *Phytocrene gigantea* in Wallich's *Plantae Asiaticae Rariores*, but (a noteworthy indication of his interest in Cryptogams at this time) he had supplied an account of the structure and development of *Targionia hypophylla* to be appended to Mirbel's classic monograph on the anatomy and physiology of *Marchantia polymorpha*—published in 1832.

His medical studies finished, Griffith sailed from England in May 1832, he arrived at Madras in September and was appointed Assistant-Surgeon on the Madras establishment in the service of the East India Company. His scientific work was done in the intervals of a busy life. Only a man of great energy and enthusiasm and possessed of great powers of physical endurance could have done the work that Griffith crowded into the 12½ years, between his landing in India and his death at Malacca before the age of 35 on February 9, 1845. This time was all spent in the East Indies—he never returned to England.

Deferring for the moment consideration of his scientific work we may take a general survey of Griffith's movements during his working life and of his labours as an explorer and collector.

After spending some months in the neighbourhood of Madras, he was situated for more than two years at Mergui and collected extensively in Tenasserim. He was recalled to Calcutta in 1835 and attached to the Bengal Presidency in order to be sent with Dr Wallich and Mr M'Clelland to visit and inspect the localities in which tea grew wild in Assam. Griffith's full report on this enquiry led to the important economic conclusion (based largely on a critical comparison of the Assam flora with the flora of tea-growing regions of China) that tea might be successfully grown under the conditions in Assam and similar districts of India. When the other members of the expedition returned Griffith was detained in Assam, where he remained during the whole of 1836, making a successful expedition into the Mishmee mountains only once before visited by a European.

Early in 1837 Griffith, accompanied by only one servant, set off on an exploring expedition through the very disturbed country of Burmah towards Rangoon. All news of him ceased, or rather his assassination was credited by the Government and reported in the newspapers, when in June he re-appeared, ragged and travel stained, in Calcutta. He had explored down the Hookhoom (Hokong) Valley and on to Ava, and had then proceeded more rapidly by river to Rangoon, conveying his collections with danger and difficulty.

Appointed Surgeon to the embassy about to start for Bhutan, he filled up the intervening two months by again going to the Khasi hills to collect. He then accompanied the expedition to Bhutan, traversing over four hundred miles of the country and returning to Calcutta in June 1838. Here he spent the next few months arranging his collections and also studying the plants of the suburbs.

In November he joined the army of the Indus and accompanied it in its whole march. He remained another year in Afghanistan making various expeditions in the country and into the Hindoo Koosh. He returned, after visiting Simla and the Nerbudda, to Calcutta in the middle of 1841.

Griffith then proceeded to Malacca where he had been appointed Civil Assistant-Surgeon. He remained only a year, but long enough to appreciate the great interest of the district for his botanical work and to complete some important observations. He collected the plants of the province and also visited Mount Ophir.

Recalled to Calcutta, he took charge of the Botanic Gardens and also lectured to the medical students during Wallich's absence from August 1842 to August 1844, pressing forward reforms in the gardens and using his opportunity for scientific observation. On Wallich's return Griffith remained for some months longer in Calcutta continuing his work, married in September, and returned to Malacca in December full of hopeful plans for scientific work there. He had barely arrived at Malacca and begun work than he was seized with a fatal illness and died on February 9, 1845.

It has been necessary to consider in some detail the rapid movements of Griffith's life in the East in order to fully appreciate the difficulties under which his large amount of scientific work was accomplished. The twelve years of his official life were filled with professional duties, difficult and dangerous exploration, management of the Botanic Gardens, and the labours entailed in making and caring for extensive collections. It would not have been surprising had Griffith, in spite of his attainments, contributed nothing to scientific botany beyond rendering these collections available for other workers. He estimated his collection of plants at more than twelve thousand species; and on his travels he did not neglect other collections of interest. Insects obtained by him are described, he collected the birds and fish in every district he visited; indeed he was a keen fisherman and must have thrown a fly in many a stream that had not been fished before, combining sport and science.

Griffith's collections were made with the definite purpose of enabling him, when he had leisure, to produce a general account of the Indian flora on a geographical basis. His methods of collecting were most enlightened and subserved his ' work as a morphologist and a student of the conditions of occurrence of the plants, not merely of formal systematic botany. The journals he kept on all expeditions are full of references to the occurrence of the plants met with. He often adopted a plan of roughly mapping each day's route and indicating the plants and associations of plants, along the line of march. I wonder if modern ecologists know of these records made long before ecology was invented?

Whenever possible he seems to have examined the morphology of the living plants, and he fully realised the value of preserving portions of the plants in spirit for future examination instead of relying on herbarium material.

This quotation from a letter to Wight (then Superintending Surgeon of the Madras Service), with whom Griffith kept up a most interesting and friendly correspondence, from which I should like to quote largely, may give an idea of his point of view and also show how he looked forward to returning to Malacca :—

"If ever you go to the place of Podostemon endeavour to get some germinating or at least very young plants. I can fancy how an Acotyledonous plant gets a stem but how a Dicotyledonous plant loses it, and becomes as some of them do, mere discs spread

over rocks is another thing. Then again where are their roots?
How opposed to late ideas of the absolute distinction of the
three great divisions. Also please to take a bottle of spirits,
and deposit specimens in it. I shall not be very sorry to get
back to Malacca, this is a delightful place truly, but one is
interrupted, and the lectures at the Medical College consume
much time. For botany no place can exceed Malacca."

And again,

"What a business it will be to settle the types of the families
from which the names must eventually be taken; this will
never be done by dried-plant botanists; but by examination
of development, which I am convinced will alone give the key."

As to Griffith's methods of work, we learn from a memorial
notice of him by Mr M'Clelland that whenever possible after
the business of the morning was finished the rest of the day
was devoted "to the examination and dissection of plants under
the microscope, drawing and describing all peculiarities presented." "Even on his death-bed his microscope stood beside
him with the unfinished drawings and papers and dissections
of plants on which he was engaged the day on which the fatal
symptoms of his disorder came on."

All his work shows the same characters of direct individual
observation and interpretation of the facts before him, repeated
examination of the same point, and almost a prodigality of
labour in recording his observations in drawings. At first under
the influence of Robert Brown, he used the simple microscope
with triplet lenses, but later he employed the compound microscope and in the year before his death writes hopefully of
ordering a first-rate microscope when he obtains the arrears
due to him from the Directors.

Griffith's high attainments were appreciated by the distinguished circle of English botanists of his time with whom he
corresponded. Mr Solby, to whom he always sent home his
papers for submission to the Linnean Society; Robert Brown,
to whose work he constantly recurs with admiration, and whose
judgment he trusted absolutely; Lindley; Sir William Hooker,
who looked forward to his being settled permanently in charge
of the Calcutta gardens, and Dr Wight may be named.

I may quote from a letter addressed to Griffith by von Martius of Munich, since it couples his own opinion and that of Robert Brown. "He (Brown) agrees with me in appreciating your spirited and enlightened investigations, and I now more than ever look forward to you as his successor—as the standard English botanist."

Only an outline of the nature of Griffith's scientific work with some details on selected subjects can be attempted here. His published works in the *Transactions* of the Linnean Society and elsewhere, important as they are, represent only a small fraction of his observations. But the wisdom and liberality of the East India Company has put us in possession of his unpublished notes and drawings (bequeathed with his collections to the Company) in the posthumously published volumes of *Notulae ad plantas Asiaticas* with the accompanying sets of plates. Though his papers were not ready or intended for publication in this form and suffer from having had to be arranged by another hand, they afford, together with his published work, a particularly good picture of how the problems of morphology and classification presented themselves to a keen investigator at this time.

Of his purely systematic work I shall not speak at length. In addition to smaller papers the most important contribution was his illustrated monograph on the Palms of British East India. In the *Notulae* numerous species are described and figured nearly always with reference to the morphology and physiology of the parts concerned. It is his investigations made with direct reference to morphology and reproduction that claim our attention most. In dealing with them it is convenient to treat of the main questions to which he directed his attention rather than of the separate papers. I shall call attention first to his work on *the flower and on fertilisation* in a number of plants, then to his *observations on Cycas*, and lastly to *his work on the Cryptogams*.

Interest in the structure of the ovule and the nature of fertilisation was widespread at the time Griffith worked. A few years previously Robert Brown had laid the foundations of the scientific study of the ovule and the behaviour of the pollen

tube, and during Griffith's time the papers of Schleiden, which
extended the comparative study of the ovule and advanced the
important though erroneous view that the embryo originated
inside the embryo-sac from the tip of the entering pollen tube,
were appearing. Schleiden's text-book did not appear until too
late to be known to Griffith. His interest was keen on continuing the work, that Brown had begun, on plants that only
a resident in the tropics had the opportunity of studying
properly, and the first volume of the *Notulae*, with the accompanying Icones, and the more systematic volume on the
Monocotyledons and Dicotyledons contain his unpublished
observations on the ovule and flowers of many plants.

His first paper in the Linnean Transactions was on the ovule
of *Santalum*. Griffith observed and rightly interpreted the free
prolongation of the embryo-sac from the nucellus, and described the application of the pollen tube to the summit of the
embryo-sac, the development of the endosperm, and the origin
and development of the embryo. He also recognised and
figured the great prolongation backwards of the embryo-sac as
an empty, absorbent caecum. At first he left the origin of the
embryo doubtful, while recognising the advantages of the exposed embryo-sac for settling the question, but later he decided
in favour of Schleiden's erroneous view that the embryo developed
from the tip of the pollen tube. Griffith also examined the
ovules of *Osyris* recognising the corresponding facts.

Comparison with the figures of Santalaceous ovules in
Guignard's later work will serve to show both the magnificent
accuracy in observation of Griffith and the limitation, running
through all the work of the time, of not recognising the contents
of the embryo-sac before fertilisation.

The Loranthaceae was another family on which the development of the embryo-sac and the processes of fertilisation and
development of the fruit interested Griffith specially. Not only
did he send his results home to the Linnean Society in two
papers, but his descriptions and figures of all the species described in the *Notulae* take account of these morphological and
developmental facts. He traced the development of the cavity
of the ovary and regarded the ovules as reduced to their

simplest expression—to an "amnios" or embryo-sac. And he observed the extension of the embryo-sacs up the style and the union of the pollen tube with the tip of the embryo-sac. His further description of the development of the embryo, endosperm and fruit is wonderfully exact if we allow for his regarding the long suspensor bearing the embryo as derived from the pollen tube growing down through the long embryo-sac.

Griffith thus recognised all the main peculiarities of *Viscum* and of *Loranthus* subsequently described more in detail in European species by Hofmeister (whose analysis of Griffith's work in 1859 is a great testimony to its accuracy) and later by Treub in the tropical species which had been studied by Griffith.

The Balanophoraceae was another group, on which Griffith made pioneer investigations. He collected and examined all the species he met with, partly from the systematic interest in supporting Robert Brown's objection to Lindley's class of Rhizantheae, but still more from his interest in the details of their reproduction. An examination of the plates from his memoirs, only published after his death, in the Linnean *Transactions* will show how fully he was aware of the structure of the archegonium-like female flower of *Balanophora* ; of the relation of the pollen-grains and pollen tubes to it; and of the appearance of the endosperm which he mistook for the embryo. Throughout he compares the structure with the pistillum (archegonium) of Bryophyta.

Thus in the Balanophoraceae also Griffith laid the foundations on which the work of Hofmeister, and more recently that of Treub and Lotsy follow.

When at Malacca Griffith interested himself among many other problems in the ovule and the development of the seed of *Avicennia*. He had previously paid attention to the viviparous embryos of other Mangroves. This piece of work, when compared with Treub's re-examination of *Avicennia*, brings out so clearly Griffith's accuracy, so far as his means of observation allowed him to go, that we may look for a moment at how these two investigations, separated by forty years, compare.

Griffith recognised the development of the embryo-sac in

the nucellus of the ovule which he took to be naked, missing the very slightly indicated integument. He followed the pollen tube to the tip of the embryo-sac and the development of the endosperm in its upper portion, where the embryo appeared. He saw the growth of the endosperm leading to its complete protrusion from the ovule and inverting the embryo so that its cotyledons point to the surface. Further he saw the long, empty, absorbent caecum grow out from the hinder end of the embryo-sac into the massive base of the young seed.

This account is substantially correct in all its facts, and Treub's work adds to it the cellular details of the origin of the embryo-sac, the setting apart of the endosperm cell to grow into the haustorium, and the details of segmentation of the embryo.

Such vivid, accurate, description of strange facts, when previous knowledge gave no clue, is in itself no mean scientific achievement.

To sum up Griffith's work on the morphology of the reproductive organs of the Angiosperms we see that he added many important facts and gave correct descriptions of what still remain among the most anomalous ovules and embryos. His methods did not enable him to distinguish clearly the contents of the embryo-sac, and he accepted and confirmed Schleiden's erroneous view of the origin of the embryo. But this hardly detracts from the directness and consequent value of all his observations.

Turning now to the Gymnosperms, we find again that Griffith devoted much attention to those forms that from his residence in the tropics he was in a position to study with most advantage. He describes in the *Notulae* his observations on the ovules and pollination of various Coniferae and Gnetaceae. But we may concentrate our interest on his work on *Cycas*. The rough structure of the young seed had already been described by Robert Brown who had recognised the gymnospermy of the group.

But Griffith's descriptions and figures are much more accurate —are indeed far in advance of those of much later observers— and add greatly to our knowledge of this plant. These two figures (pl. XVI) will speak for themselves and show how clearly Griffith

had grasped the morphology of the Cycadean ovule, how faithfully he delineated the details, and how he sought in progressive
development to throw light on the structure.   He added to the
previously imperfect description of the ovule an accurate account
of the pollen chamber, and the proof that pollen grains entered
and filled it.   Further he followed the germination of the pollen
grains, not merely recording the fact that the tubes penetrated
the nucellus all around the pollen chamber, but ascertaining
in how many days the tubes were put forth.   His fullest description is unfortunately displaced in the *Notulae* under the
heading of *Thuja*, but it is clear that it refers to the *Cycas*
figured on the same plate as that plant.

From what has been said of the nature of Griffith's work on
the ovules, both of Angiosperms and Gymnosperms, the complete
omission of his name in recent works on the two groups that
are in constant use is at least noteworthy.

Griffith was specially interested in the study of Cryptogamic
plants.   In a letter to Wight he says "I would like to be out
with a work on Indian Cryptogamia of higher forms; so much
so that if I see no chance of my succeeding to the Gardens, I
intend sending away all my other collections, and devoting
myself to this object and general development, which is
obviously the keystone of the arch."

He left Algae and Fungi (with the exception of the Characeae)
alone, and it is his work on the Bryophyta and Pteridophyta that
concerns us.   For information on his views on these plants we
are dependent on his paper on *Salvinia* and *Azolla* and on the
Notulae, put together as I have said from his notes after his
death, and not intended for publication in this form.   But there
is no difficulty in getting a clear grasp of his point of view.
This was a mistaken one—an attempt to bring into line the
reproduction of the gametophyte of Bryophytes, the sporophyte
of Vascular Cryptogams, and the flowering plant with its flower
and fruit.   It is easy to be wise after the event.   In these
comparisons Griffith belonged to his time with a much wider
field of personal observation than most possessed.

We must bear in mind that at the time when Griffith worked
no idea of the sexual and asexual alternating generations in

Plate XVI

From Griffith's *Notulae*

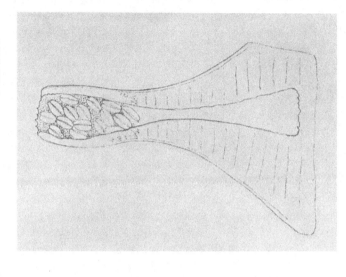

Nucellar apex of Cycas with pollen chamber
and pollen grains

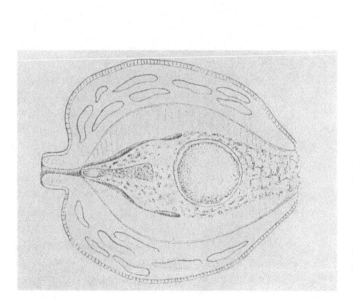

Median section of the ovule of Cycas

Pteridophytes had been gained, although the prothallia had been observed preceding the growth of the plant in *Equisetum* and Ferns. It was not till some years after Griffith's death that fuller facts as to the sexual organs were obtained and led to the right comparisons.

Griffith's work on the Bryophyta shows the same power of observation as that on the ovule, but the difficulties due to imperfect instruments are more evident. His views on reproduction were here, however, clear, since the development of the capsule was definitely related to the fertilisation of the pistilla (archegonia) by the substance formed in the anthers. His figures indicate how much he saw, and how here also he sought in development the interpretation of mature structure.

His early interest in the Liverworts, especially the Marchantiaceae, continued, and all the forms he collected were carefully examined and figured with his usual accuracy.

One of the Liverworts Griffith described may be taken as an illustration to this part of our subject on account of the interest of its re-discovery and re-description in 1910 by Goebel. This is a plant collected in Assam and named *Monosolenium tenerum*. This Marchantiaceous plant is described as having no air-chamber layer, as bearing sessile, dorsal, antheridial receptacles, and terminal, shortly stalked archegoniophores with one ventral groove in the stalk. A single archegonium—later capsule—is found in each of the half-dozen involucres. Spores and irregular bodies were found in the capsule.

Recently Goebel had two tea-plants sent home from Canton. They died, but he kept the soil moist on the chance of germinating seeds. Among a number of other plants there turned up a new Liverwort. On examination this proved to be Griffith's *Monosolenium*—all types of which had been lost—a most interesting form related to the Corsiniaceae.

In the Mosses and the Liverworts generally Griffith was clear on the development of the capsule or fruit following on the impregnation of an archegonium. But in *Anthoceros*, while he recognised the antheridia he was not clear as to the sunken archegonia, and regarded the capsules as arising by impregnation of unrecognisable spots on the young frond or thallus. He observed however the indication of the canal of the archegonial neck above the young capsule.

Analogy with *Anthoceros* confirmed him in his views on the reproduction of ferns. Here he spent much labour in considering the view, originally due to Hedwig, that the ramenta were male organs by the effect of which the sporangia developed. Griffith saw that if this was so, since the sporangia are initiated very early, the only time to search for the male organs was in the very young stage of the leaf. On examining such young leaves he found the terminal cells of the young ramenta very prominent and formed the working hypothesis that they were the male organs. But he stated this cautiously and was well aware how imperfect his means of observation were.

The whole line of work brings vividly before us how cryptogamic the Cryptogams were at this period.

Without attempting to survey Griffith's views on the various groups of Vascular Cryptogams, a word must be said of those on *Salvinia* and *Azolla*, on which he published a long paper in addition to the other descriptions and figures in the *Notulae*. His observations bear on the development of the sorus and sporangium, but he dismissed the microsporangia as abortive or imperfectly developed structures. (I may note in passing that the study of their development led him to regard the microsporangia of *Isoetes* in the same way.) He dwelt on the similarity of the sporangium and indusium of *Azolla* to a gymnospermus ovule, and regarded the filaments of *Anabena* seen penetrating within the indusium as probably the fertilising bodies in this naked-seeded cryptogam.

Thus with a large amount of fresh and original observation Griffith was on wrong lines in his general views and comparison —he classed the higher Cryptogams in his *Notulae* as

Pistilligerous.   Musci. Hepaticae.

Gymnospermous. Azolla. Salvinia. Chara.

Cryptogamous.   Ferns. Lycopods. Isoetes. Marsilidae.
                    Anthocerotidae. Equisetidae.

Griffith's general views of the reproduction of all the Vascular Cryptogams was necessarily wrong, since the prime clue of the recognition of the prothallus and plant as distinct had not been

found. In this connection his figuring young plants of *Equisetum*
attached to prothalli is interesting. In some speculations concerning the embryology of *Loranthus* he came, by a wrong
line of approach, within touch of the right comparison, when
he compares the endosperm to the confervoid green growth
(i.e. the prothallus) at the base of the young plant of *Equisetum*.

It is idle to speculate on what might have happened had
such a wide observer as Griffith chanced on the clue. In this
respect he was of his time as most are. The man who put the
industrious but blind gropings of this period in morphological
botany straight, both as regards the development of the embryo
and the comparative ontogeny of archegoniate plants was Hofmeister, and like all exceptional men he belonged to the new
period created by him.

The great advantage of this course of lectures seems to me
to be that it approaches the study of the history of botany in
the right way ; for progress in our science has been the result
of individuals rather than of schools. The consideration of the
work of Griffith from 1832 to 1845 is a vivid illustration of the
condition of morphological botany in the earlier portion of the
period, surveyed in one of the chapters in Sach's *History* under
the title of "Morphology and Systematic Botany under the
influence of the History of Development and the knowledge of
the Cryptogams." These two subjects were always before
Griffith.

The interest of the personality of William Griffith and of
the work he accomplished in his tragically short life is obvious.
Not less so is the way in which that work was done inside the
limitations of his period. We, who are still gleaners in the field
that Griffith and his contemporaries cleared and Hofmeister
marked out and tilled, are probably just as incapable of con-
ceiving the future developments of morphology.

ARTHUR HENFREY
1819—1859

By F. W. OLIVER

Narrative—state of Botany—dawn of the Golden Age—sexuality of Angio-sperms—Schleiden's elucidation of fern life-history—Nägeli, Suminski and Hofmeister—recognition by Henfrey—original work—publications —the *Micrographic Dictionary*—*The Botanical Gazette*—its features— Henfrey's labours not immediately productive.

THE claim of Henfrey to rank among the founders of botany in this country depends less on his own original contributions than on a whole-hearted devotion to the propagation and diffusion of the newer methods and results which marked an epoch during the forties and fifties of last century. The outset of Henfrey's career coincided with a great turning point in the history of botany, and to Henfrey will always belong the credit of being the first Englishman to recognise the full significance of the movement. From that moment he unceasingly made known and diffused in this country the results of the German renaissance. That Henfrey should have failed to establish the newer botany in England was the result of a variety of circumstances, one of which was his early death.

The available biographical material of Henfrey being ex-tremely meagre, it has been necessary in preparing the present account to rely almost entirely on his published writings. In some ways this lack of personal details is no disadvantage as our present interest in Henfrey depends essentially on the movement in botany with which he was identified.

Arthur Henfrey was born at Aberdeen, in 1819, of English parents. He underwent the usual course of training for the medical profession at St Bartholomew's Hospital—becoming a member of the Royal College of Surgeons in 1843. In consequence of bronchial trouble, to which he eventually succumbed at the early age of 39, Henfrey never practised his profession. Compelled to a life of seclusion he at once turned to a scientific career and more particularly to the pursuit of botany. In 1847 he undertook the duties of Lecturer in Botany at St George's Hospital Medical School, where among his colleagues was Edwin Lankester, himself a redoubtable naturalist and the father of Sir Ray Lankester, the eminent zoologist of our own day.

Henfrey succeeded Edward Forbes as Professor of Botany in King's College, London, in 1852—a post which he held till his death. He was elected to the Fellowship of the Royal Society in the same year.

He died quite suddenly in 1859, at the house on Turnham Green, where he had resided for many years.

In order to understand the part played by Henfrey, it is necessary briefly to review the state of botany in the first half of the nineteenth century.

Linnaeus of course, botanically, the outstanding fact of the eighteenth century, was no exception to the dictum that "the evil that men do lives after them."

It was supposed that botany had reached its culminating point in Linnaeus and that improvement could only be made in details—elaborating and extending his system. As Sachs tells us in his *History*, the result was that "Botany ceased to be a science; even the describing of plants which Linnaeus had raised to an art became once more loose and negligent in the hands of his successors. Botany gradually degenerated under the influence of his authority into an insipid dilettantism—a dull occupation for plant collectors who called themselves systematists, in entire contravention of the meaning of the word."

This was written with especial reference to Germany, but it applied with no less force to our own country where the Linnaean idea had taken deep root and the Linnaean collections had found a sanctuary.

However, by 1840, a change was coming over the face of botany. Little as it can have been dreamt, the Golden Age was already beginning—destined in a relatively short time to transform the subject. This Golden Age was contemporaneous with, and immediately dependent on, the rise of a group of young botanists in the Fatherland, a group which included von Mohl, Schleiden, Hofmeister, Nägeli, Cohn and De Bary. Later it was reinforced by Sachs, who in addition to being a brilliant physiologist was a gifted writer who did much to establish scientific botany on a sound footing. It is impossible to overestimate the debt due to Sachs, particularly for his great *Textbook of Botany*, which at the right psychological moment brought the whole of the modern work between the covers of a single volume.

It was with the dawn of this period that Henfrey identified himself. In the 15 years of his active career (1844—1859) he devoted himself very largely to making his fellow-countrymen acquainted with the newer aspects of botany. More particularly it was the recent discoveries as to the reproduction and life history of the Vascular Cryptogams that specially engaged his interest— the researches which broadly speaking we associate with Hofmeister to-day.

Before we go on to speak of the sexuality of the Cryptogams however, a few words may be devoted to that of the flowering plants.

Sexuality of Flowering Plants. At the period when Henfrey entered on his career as a botanist no reasonable doubt remained as to the existence of sexes among the flowering plants. The theory of the sexual significance of the organs of the flower, brilliantly founded by Koelreuter in the previous century, had been perfected with a great volume of experimental proof by K. F. Gaertner the son of Joseph Gaertner of *Carpologia* fame.

By 1830 the mechanism of fertilisation came to light in Amici's discovery of the pollen tube which he traced from the stigma to the micropyle. The microscopic aspect of the problem was taken up with great energy by Schleiden and brought to the forefront as the burning question of the early forties. The theory of Schleiden, which applied in particular to the flowering plants, made its influence felt to such an extent in the search for evidence of sexuality among the Cryptogams, that we may conveniently state in a few words in what it consisted.

Schleiden traced the pollen tube into the micropyle, and

thence to the nucellus where it depressed or invaginated the apex of the embryo-sac, and in the recess or indentation so produced the tip of the pollen tube was converted into the embryo—its actual apex being represented by the plumule. This theory was the lineal descendant in modernised trappings of the old view expressed by Morland and others at the beginning of the eighteenth century that the embryo was contained in the pollen grain, and that the ovule was no more than the brood chamber whither it must be brought to undergo further development. This erroneous interpretation of the true facts was always repudiated by Amici, and was finally overthrown by Hofmeister and Radlkofer in the early fifties. In this connection we may note in passing Henfrey's careful paper on the impregnation of *Orchis Morio*, published in 1856, which fully corroborated Amici. In this paper the relations of pollen tube, embryo-sac, egg-cell, suspensor and embryo were correctly interpreted, and the new point established, contrary to the assertions of previous observers, that the ovum or "germinal-vesicle," prior to fertilisation, was a naked, unwalled cell.

Sexuality in Cryptogams. By far the most important question that came to a head in Henfrey's time was that of the morphological relationships of the Cryptogams and flowering plants. Hitherto these had remained altogether obscure in the absence of reliable data based on the proper application of the microscope to the elucidation of the life histories of the lower plants. Under the influence of the Linnaean school, which had taken deep root in this country, as elsewhere, the systematic study of flowering plants had been widely pursued, and in so far as the ferns were concerned their homologies were commonly interpreted in terms of the flowering plants. Without any real guidance in fact, a great diversity of views of these homologies found expression. The following, taken from Lindley, may serve to illustrate their general nature.

The sorus was regarded as a sort of compound fruit, the sporangium as a carpel, the annulus as its midrib, and the spores as the seeds. Speculations such as these are of the same order as the crude conjectures which with less excuse relieve the answer books of examination candidates at the present time.

In the search for the male organs of the fern attention was naturally directed to the neighbourhood of the sorus, and the stomata, indusia and glandular appendages were in turn mistaken by various observers for the anthers. The "limit" was reached by Griffith who, as is stated at page 190, conjectured that the Anabena filaments which accompany the megasporangia of Azolla were no other than the male organs of that plant. Schleiden spoke of these researches with the utmost scorn. "For my part I am surprised that no one has yet insisted upon the presence of the organs of sense, as eyes and ears in plants, since they are possessed by animals. Such an assumption would not be a bit more absurd than the mania of insisting upon having anthers in the Cryptogams, simply because they are found in the Phanerogams."

All these ill-grounded hypotheses were swept away in 1844 when Nägeli discovered antheridia containing spermatozoids on the "cotyledon" or pro-embryo of the fern—the prothallus we call it now. Nägeli at once recognised their essential agreement with the antheridia already known in the Bryophytes and compared the spermatozoids with the corresponding structures in animals. But as he overlooked the existence of the archegonia, or rather by some lapse mistook them for stages in the development of the antheridia, it is not surprising that he was at a loss to understand the significance of his discovery, and that he should have commented on his dilemma in the following terms. "Seeing that the female organs (spores) arise on the frond at a much later stage of development, and long after the pro-embryo has died away, the function of the spermatozoids is far from evident."

It was only three years later that light was thrown on the situation, and from an unexpected quarter. Count Suminski, an amateur microscopist, announced the discovery of additional reproductive organs on the fern pro-embryo, which he clearly distinguished from the "spiral filament organs" or antheridia. His full paper, which appeared in 1848, marks an epoch in morphology, and was a very remarkable performance. In it he redescribes the antheridia and spermatozoids—detecting their tufted cilia which Nägeli had overlooked. The archegonia he describes as ovules without envelopes consisting of a·papilla (the neck) which becomes perforated, giving the spermatozoid access to the embryo-sac within. His figures of the process of fertilisation are extremely interesting as they show how completely he was dominated by the theory of Schleiden to which allusion has already been made. The head of the sperm is represented as entering the "embryo-sac," and there becoming encysted to form the embryo just as the tip of the pollen tube was supposed to do in flowering plants. The further development of the embryo and its various organs are traced and figured, however, in the most admirable way. At the conclusion of his paper Suminski states that in view of the presence of male organs and ovules, and the occurrence of fertilisation, the cryptogamy of ferns does not exist in a physiological sense, and ceases to have any validity as a peculiar character. A remark which he follows up by the statement that ferns must on the existing classification be referred to the Monocotyledons.

In certain respects no doubt Suminski's paper is fantastic— more especially the circumstantial details given of the process of fertilisation. But, however we may criticise his work the credit belongs to Suminski of showing (1) that sexual organs are borne on the prothallus, (2) that the embryo fern plant is produced as the result of fertilisation. Unlike Nägeli, to Suminski came the happy inspiration of looking for the female organs in the position where common sense indicated they ought to be found.

Suminski's paper instantly aroused universal interest, and the whole of his assertions were at first categorically denied by the German botanist Wigand.

We may now trace Henfrey's attitude to Suminski's work.

His first notice occurs in the body of a review of Lindley's "Introduction" in the first volume of his *Botanical Gazette*, and shows him to have been profoundly sceptical, if not contemptuous, of the occurrence of fertilisation in the prothallus of the fern. His words are "this (i.e. Suminski's discovery) appears to have little but originality to render it worthy of notice." That appeared in February 1849.

Writing at greater length of Suminski's work in the *Annals*

and Magazine of Natural History, in November of the same
year, he speaks much more guardedly. "These researches are
in the highest degree curious, and if the facts related prove to be
correct, most importantly affect the received views of analogies
in the generative processes of plants."

At the same time Henfrey says he hopes to speak more
definitely on this matter when his own investigations are complete. Two years later his own very careful work in the same
field was laid before the Linnean Society, in which he corroborated the main facts that had come to light. Turning once
again to the paper of Suminski, after making certain criticisms
of detail, Henfrey handsomely remarks—"Nothing however can
take from him the credit of having discovered the *archegonia*
and their import, one of the most important discoveries in
physiological Botany of modern times since it has led to results
revolutionising the whole theory of the reproduction of plants
and opened out a totally new sphere of inquiry into the laws
and relations of vegetable life."

For some little time after these discoveries the archegonia of
the fern were, on the initiative of Mercklin, commonly referred to
as the "organs of Suminski," a custom which happily fell into
desuetude. Mercklin, in his paper, which essentially repeats the
work of Suminski, states that he devoted his entire attention
for three months to the fern prothalli before he succeeded in
observing the entrance of a spermatozoid.

In reviewing the early papers of the Hofmeisterian epoch—
papers which form the bed-rock of the existing morphology—
one is struck with the marvellous rapidity with which their significance was apprehended. We find the phrase "alternation of
generations" employed within two years of the discoveries of
Suminski, whilst by the early fifties the general genetic relations
of the vascular series were realized in quite a new light.

As Sachs puts it :—"When Darwin's theory was given to the
world eight years after Hofmeister's investigations, the relations
of affinity between the great divisions of the vegetable kingdom
were so well established and so patent that the theory of descent
had only to accept what genetic morphology had actually brought
to view."

Among Henfrey's original contributions other than those dealing with the burning questions already mentioned, was a series dealing with the Anatomy of Monocotyledons. This would appear to have led him on to study the Nymphaeaceae, and especially the anatomy of *Victoria regia*—a paper which may be compared perhaps with Prof. Gwynne-Vaughan's more recent study. Henfrey was quite alive to the monocotyledonous affinity, and the enlightened and, for that date, unconventional views to which he gave expression, drew an interesting notice by Hooker and Thomson in the first volume of their *Indian Flora*.

Another of his papers dealt rather fully with the development of the spores and elaters of Marchantia, where he filled in a considerable lacuna in the knowledge of that group. It is curious to find as late as 1855 so intelligent and well informed a botanist as Henfrey laying it down that the cells of Marchantia, in particular, and Liverworts in general, were destitute of nuclei. It is superfluous to say that this apprehension was quite baseless. Indeed, forty years later, the group of the Liverworts was deliberately chosen by Prof. J. B. Farmer, for the investigation of nuclear phenomena on account of the favourable conditions under which they could be studied!

Microtechnique at that time was of course a much simpler affair than it has since become. Contemporary papers as a rule say little about methods ; however one of Henfrey's occasional notes in a magazine tells us that caustic potash, iodine, sulphuric, hydrochloric and acetic acids, together with ether were in common use. Schultze's reagent—chloride of zinc iodide—was invented in 1850, but does not appear to have been generally employed till many years later.

It would however be a serious error to underestimate the value of the earlier work in plant histology. The present writer once spent an interesting morning in Pfeffer's laboratory at Tübingen rummaging through hundreds of the great von Mohl's anatomical preparations. Among these were sections of palm endosperms in which the, at that time recently discovered, continuity of the protoplasm through the cell walls was plainly visible. The existence of these filaments had been detected by von Mohl some years before, but he had refrained from publishing his observations from over-cautiousness.

As a translator and editor Henfrey was responsible for the English edition of von Mohl's *Principles of the Anatomy and Physiology of the Vegetable Cell*, published in 1852, for two volumes of Reports on Botany in the Ray Society's publications, whilst he had a considerable share in Lankester's translation of Schleiden's famous *Principles of Scientific Botany*, 1847. In addition to these there were constant abstracts and critical reviews from his pen in the *Annals and Magazine of Natural History*—a journal of which he became botanical editor before the close of his life.

As a writer of text-books Henfrey was very prolific. First came his *Outlines of Botany*, 1847, followed by the *Rudiments of Botany*. Much more ambitious was his *Elementary Course of Botany* which became a standard text-book running through numerous editions after his death, under the editorship of the late Dr M. T. Masters. To these must be added, in conjunction with Griffith[1], the *Micrographic Dictionary*, a substantial volume dealing in innumerable special and general articles with the microscopic study of plants and animals. This work was no mere compilation, but embodied in its pages is a very large amount of independent observation. The illustrations covering nearly fifty plates were by Tuffen West, and reached a high degree of excellence. A well known botanist, a contributor to the present volume, has more than once assured me that it was to the *Micrographic Dictionary* that he owed his salvation!

Should anyone desire to get a vivid and accurate picture of the precise state of Botany in this country at the middle of the last century, he cannot do better than turn over the pages of *The Botanical Gazette*, a monthly journal of the progress of British botany, founded and conducted by Henfrey. It was about the size of our own *New Phytologist*, with which it had not a little in common. In one respect it differed; unlike the *New Phytologist* the *Gazette* was financially a failure and after carrying it on at his own expense for three years (1849—1851) Henfrey had to relinquish the undertaking.

[1] Not the William Griffith of the last chapter.

A perusal of its contents clearly shows that its editor regarded his journal as one of the instruments of diffusing the New Botany. Having to rely largely for his subscribers upon the amateur collector he points out in the prefatory note that a feature will be made not only of home botany but also of contributions or abstracts from abroad dealing with floras which have much in common with our own. For the benefit of those whose collections had reached considerable dimensions, and for whom the lack of new plants might connote a waning stimulus, he held out the further inducement of papers on the general anatomy of familiar plants, of which an excellent example by Thilo Irmisch on the stolons of Epilobium was included in the first number.

For the three years of its existence Henfrey kept faith with the British botanists and a number of *The Botanical Gazette* rarely appeared without an article contrived for their edification. The task was evidently a congenial one, for Henfrey had a sound knowledge of British plants with especial reference to geographical distribution and critical forms. Unlike several later exponents of the New Botany, Henfrey was quite able to hold his own with the systematists. He more than once expresses the opinion that there was too great a tendency to lump species in the handbooks to the Flora, and he urged on the occasion of the preparation of the third edition of the *London Catalogue of British Plants* that many more forms should find recognition. The editors of the catalogue however successfully opposed the suggestion on the ingenious grounds that it would raise the weight for postage beyond the limits of a blue (twopenny) stamp!

Henfrey thought much might be done by cultivation under varying conditions to settle vexed questions as to critical species, and suggested that a limited number of botanists in different parts of the country should co-operate in a scheme under which seed should annually be distributed, harvested and re-distributed among those taking part. Henfrey himself offered to undertake the somewhat onerous duty of receiving and distributing the seed and of generally correlating the work. As however his proposal was merely tagged on to a note on *Sagina apetala* and *ciliata* it is hardly remarkable that nothing came of it.

An interesting minor feature of the *Gazette* was the reporting of the proceedings of the various Botanical Societies throughout the country. These show that a chronic state of intellectual famine frequently obtained even at the leading societies— a state of which vestiges are still occasionally discernible. It was no unusual occurrence at the Linnean even during the period of Robert Brown's presidency for the meeting to be regaled with long extracts from the commentaries on the *Hortus Malabaricus*. In this respect however the record was easily held by the now defunct Botanical Society of London, which eked out its programme for a whole year with a communication by a Mr D. Stock " On the Botany of Bungay, Suffolk." Begun on the 11th October, 1850, it only drew to a conclusion on the 3rd October, 1851. There were other attractive features in *The Botanical Gazette* on which space does not allow me to dwell.

The general impression gained, however, from a perusal of the papers of that time is that they were refreshingly short, as compared with our own day, and often very much to the point. The recording of observations was rarely made the occasion for a survey of the whole field of botany, and little trace was discernible of the present habit of over-elaboration.

The foregoing outline of Henfrey's activities shows that they were devoted wholly to the spread of the Newer Botany in this country. The means employed included the publication of reviews and abstracts, the editing of translations of the more notable books, the founding of journals, and the writing of text-books. Moreover by his own investigations he kept close touch with the modern work and was indeed the means of corroborating and often materially advancing many of the larger problems before putting them into general circulation in this country.

And yet, in spite of this complete devotion of his life to the cause, the New Botany found no permanent place in this country till twenty years after Henfrey's death.

Botanically speaking, the organisation and rise of taxonomy

was the ruling pre-occupation of the period under consideration, a direct outcome of colonial expansion and consolidation. Fed on unlimited supplies of new material from the ends of the earth the taxonomic habit became supreme. What could an isolated student and recluse like Henfrey do to stem this flood ? Circumstances were too strong for him, and founding no immediate school it remained for a later generation to take up the task.

Though the history of the establishment of the New Botany in England lies outside the province of this lecture, it is instructive, as a contrast in methods, to note the manner of its accomplishment. Henfrey, who relied on his pen, had proved ineffective to bring about a revolution. Twenty years later it fell to Sir William Thiselton-Dyer, then a young man, to succeed where Henfrey had failed. By his enlightened teaching and personal magnetism, Thiselton-Dyer aroused a widespread interest in laboratory botany. But the matter was not allowed to rest there. Holding as he did an important post at Kew, the strategic centre, he was able to obtain appointments in the chief Colleges and Universities of the country for the recruits whom he had attracted. In this way, by the exercise of an acute intelligence amounting to statesmanship, and in a very short period of time, the New Botany became everywhere firmly established.

WILLIAM HENRY HARVEY

1811—1866

By R. LLOYD PRAEGER

Early influences—Natural History—his "pretensions" to science—choice
of a profession—visit to London—early publication—South Africa—
investigation of its flora—appointed Keeper of University Herbarium,
Dublin—Algology with Mrs Griffiths—*Phycologia Britannica*—appointed
to Professorship—visit to America—lectures and travels—*Nereis Boreali-
Americana* — travels in the East — Australia — New Zealand—Fiji—
return home—election to chair at Trinity College, Dublin—*Phycologia
Australica*—marriage and death—Harvey's limitations—his reception
of Darwinism—personal characteristics.

AMONG the many illustrious names that figure on the
syllabus of the present course of lectures, that of Harvey is
probably one of the less generally known. This is due for the
most part to the fact that the subject to which the greater
portion of his energies was devoted—the systematic study of
seaweeds—occupies a somewhat remote niche in the edifice
of botany. Also many years of his life were spent in collecting
in distant regions; and his retiring disposition, and comparatively early death, contributed to the same result. In the
scientific world of his day he avoided publicity, but laboured
with indomitable zeal at his chosen subject, leaving behind
him a series of splendidly illustrated descriptive works. For
a glimpse of the man himself—his life, his aims, his thoughts—
we have to rely almost entirely on a volume[1] consisting mainly
of letters written to relations and to family friends, which was

[1] *Memoir of W. H. Harvey, M.D., F.R.S., with selections from his journal and
correspondence.* London, 1869.

Plate XVII

WILLIAM HENRY HARVEY

edited by his cousin, Mrs Lydia Fisher, and published a few years after his death. My indebtedness to this volume in what follows will be apparent.

William Henry Harvey came of the old Quaker stock that has given to Ireland several of her most enthusiastic naturalists. To this group belong Thomas Wright of Cork, Joseph Wright of Belfast, Greenwood Pim of Dublin; all of whom, immersed in affairs of business, devoted their leisure hours to science, and progressed far in the branches of zoology or botany to which they addressed themselves. Harvey's family belonged to Youghal, on the coast of Co. Cork. His father was a well-known merchant of Limerick, in which town he himself was born, the youngest of eleven children, just a hundred years ago—in February 1811. Even as a child, his love of natural history made itself apparent, and fortunately his schooling tended to foster this taste. After a few years at Newtown near Waterford, he went to the historic school of Ballitore, in the county of Kildare. These Irish Quaker schools have long favoured the teaching of science, and Ballitore at that time was no exception. The head master was James White, a keen naturalist, and himself a writer on Irish botany[1]; and probably the encouragement that young Harvey received at Ballitore had much to do with the shaping of his life. At the age of fifteen, we find him writing of his collection of butterflies and shells, and already referring to the group in which he subsequently achieved his greatest fame :—

"I also intend to study my favourite and *useless* class, *Cryptogamia.* I think I hear thee say, Tut-tut! But no matter. To be useless, various, and abstruse, is a sufficient recommendation of a science to make it pleasing to me. I don't know how I shall ever find out the different genera of mosses. Lichens I think will be easy" (he little knew them!) "but fungi I shall not attempt; not at all from their *difficulty*, but only because they are not easily preserved. But do not say that the study of *Cryptogamia* is useless. Remember that it was from the genus Fucus that iodine was discovered."

Another letter of this period, written when he was sixteen,

---

[1] An essay on the indigenous grasses of Ireland. 8vo. Dublin, 1808.

contains so quaint a description of himself that I am tempted
to quote from it :—

"In person I am tall, and in a good degree awkward. I am
silent, and when I do speak say little, particularly to people of
whom I am afraid, or with whom I am not intimate. I care
not for city sports, or for the diversions of the country. I am
equally unknown to any healthful amusement of boys. I cannot
swim nor skate. I know nothing of the delight of these, and
yet I can amuse myself and be quite happy, seemingly without
any one to share my happiness. My botanical knowledge
extends to about thirty of the commonest plants. I am very
fond of botany, but I have not much opportunity of learning
anything, because I have only to show the plant to James
White, who tells me all about it, which I forget the next
minute. My mineralogy embraces about twelve minerals, of
which I know only the names. I am totally unacquainted with
foreign shells, and know only about two hundred and fifty
native ones. As to ornithology, I have stuffed about thirteen
birds. In chemistry I read a few books, and tried some experiments. In lithography I broke a stone and a printing press.
These are my pretensions to science."

The reference to lithography is interesting, in view of the
fact that he became later on one of the most exquisite de-
lineators of plants, and with his own hand drew on stone the
greater part of the splendid plates which enrich his works on
Phanerogams and Algae. In his confession of ignorance of
sports and pastimes, we already see the result of the want of
robust health which followed him through life, and brought
about his premature death ; and in spite of which he performed
such monumental work.

Already Harvey's mind was quite made up as to what line
in life he would prefer. He cannot hope, he says, to achieve
success in commerce, by "buying cheap and selling dear." As
regards professions, he is "neither fit to be a doctor nor a
lawyer, lacking courage for the one, and face for the other, and
application for both....All I have a taste for is natural history,
and that might possibly lead in days to come to a genus called
Harveya, and the letters F.L.S. after my name, and with that

I shall be content....The utmost extent of my ambition would be to get a professorship of natural history."

His parents had thought of placing the boy with an eminent chemist in London, but his obvious antipathy to the prospect of city life led to his entering his father's office in Limerick instead. The quiet home life which ensued was well suited to his taste. All holidays were devoted to collecting. The family had a summer residence at Miltown Malbay, on the Atlantic coast, an excellent spot for Algae; and it was no doubt the time spent there that brought these plants prominently under his notice, and led to the noteworthy researches of later days. For the time, Mollusca still mainly occupied his mind, and in 1829, at eighteen years of age, we find him busily engaged in drawing the plates for a *Testacea Hibernica*—a book that never saw the light, though two years later he writes of being at work on his *Bivalvia Hibernica*, which was then half finished.

In the same year, he made his first excursion into "foreign parts" as he calls them, visiting Dublin, Liverpool, London, Edinburgh and Glasgow. An account of a meeting of the Linnean Society, to which he was taken by his friend Bicheno, then secretary, and at which "if not edified I was amused," shows that the reverence he felt for science did not necessarily extend to constituted scientific authority. "The President wore a three-cocked hat of ample dimensions, and sat in a crimson arm-chair in great state. I saw a number of new Fellows admitted. They were marched one by one to the president, who rose, and taking them by the hand, admitted them. The process costs £25."

In 1831, his finding at Killarney of the beautiful moss *Hookeria laetevirens*, hitherto unknown in Ireland, led to the formation of one of the warmest and most valuable friendships of his life. He forwarded specimens, with a characteristic letter, to W. J. Hooker at Glasgow, and the kind and encouraging reply which he received led to further letters and eventually to an intimacy which seems to have been prized equally on both sides. Hooker recognized at once the extraordinary talent of the shy young man of twenty, lent him books, asked him to visit him, and congratulated him on his critical faculty,

predicting for him a rapid advance to "the top of algologists."
Another life-long friendship made about this time was with
Mrs Griffiths of Torquay; and he numbered Greville and
Agardh among his earliest correspondents. Already he was
deep in his life-task of comparing and describing plants,
working with the restless energy which characterised him.
"I rise at five every morning," he writes, "and work till breakfast, examining or describing the Algae for the 'British Flora[1].'
If I do five species a day I think it good work. This may
seem slow, but there is much to be compared and *corrected*! for
I differ from Dr Hooker on many species. Oh, impudence!
oh, presumption!" In 1832 he undertook to do the Algae for
J. T. Mackay's *Flora Hibernica*, which was published three years
later; this was his most important contribution to the botany of
his native land.

The death of his father in 1834 broke up Harvey's home life,
and his strong desire to study the vegetation of distant countries
led to enquiries as to the obtaining of an appointment in the
Colonies. New South Wales was first thought of, but it was for
the Cape that he started in the following year.

Asa Gray, a friend of many years' standing, tells, in a notice
of Harvey in the *American Journal of Science and Arts*[2], a curious
story as to the circumstances attending this momentous change
in Harvey's life. The story is repeated in the notice of Harvey
in Seemann's *Journal of Botany*[3], though not mentioned in the
Memoir edited by his cousin. It seems that, as the result of
Harvey's representations, he obtained through Mr Spring Rice,
afterwards Lord Monteagle, the post of Treasurer at the Cape;
but, by an accident, the appointment was made out in the name
of an elder brother (Joseph Harvey); and an inopportune change
of ministry occurring just at the time, frustrated all attempts at
rectification. Be that as it may, Joseph Harvey sailed for South
Africa in July 1835, taking his younger brother with him as
assistant.

It was with high hopes that the naturalist started for the
Southern Hemisphere. At that time the flora of South Africa

[1] Published as Vol. v., Part 1, of Smith's *English Flora*.
[2] Vol. XLII. p. 274, 1866.   [3] Vol. IV. p. 236, 1866.

was but slightly known. About Cape Town itself and near other older centres of colonization, indeed, many plants had been collected, both by Dutch and English; but vast tracts of mountain and veldt, for a thousand miles to north and east, were still unexplored. He describes his excitement on landing, and how, after a sleepless night, he started off for the hills early next morning, to revel among strange Ericas, Polygalas, Lobelias, Diosmas, Proteas, and Ixias. He at once settled down to collecting with his usual method and energy. From four or five until nine every morning he was at work on the mountains or on the shore; after which several hours were devoted to preserving the material. Within a few weeks he was engaged on the description of new genera and species, and in three months his herbarium contained 800 species. Already schemes for organized work leading up to publication were in his mind; and it seemed as if his task lay open before him; but fate willed otherwise. His brother fell ill within a few months of his arrival, and a little later a return to Europe was ordered—to no purpose, as Joseph Harvey died on 26 April, a fortnight after sailing, and it was a sad home-coming which the naturalist, who had accompanied the invalid, experienced in the June following. He started again for South Africa a few weeks later, to take up his brother's duties as Colonial Treasurer; and remained there for three years, when severe illness, brought on by overwork, compelled a return home. But he came back, and resumed his strenuous life, spending his days in official duties and his nights at botany, until, in 1842, a complete break-down forced him to resign his post, and leave the country. Seven years of his life were thus devoted to South Africa, and, in spite of the serious inroads on his time and energy caused by two tedious voyages home, as well as by illness when at the Cape, a great amount of botanical work was accomplished. He arranged with collectors for the supply of plants from various parts of the country; he got the Government interested in the native flora, so that official papers were issued giving instructions for collecting and soliciting specimens; and Harvey himself devoted so much time to his hobby that he suggests that his title should be Her Majesty's Pleasurer-General, instead of Treasurer-General. Every month brought its quota of undescribed plants. "Almost every small package of specimens received from the Natal, or the Transvaal district," he writes[1], "contains not only new species, but new genera; and some of the latter are of so marked and isolated a character, as to lead us to infer in the same region the existence of unknown types that may better connect them with Genera or Orders already known." To produce system in this chaos he compiled and published his *Genera of South African Plants* (1838), the forerunner of the larger works which constitute his principal memorial in the domain of Phanerogamic Botany. But the uncongenial climate and the intense application were too great a strain on his health and he reached Europe in 1842, prostrated in both body and mind.

Nevertheless, the final year of his residence in Africa saw the production of the first of the series of works on seaweeds by which his name will ever be best known. His *Manual of British Algae* was issued by the Ray Society in 1841, its Introduction dated at Cape Town, October 1840—a modest octavo volume, characterized by the thoroughness which runs through all his work.

A period of convalescence and apathy followed his return, in which he wandered about Ireland, doing some desultory botanizing; after which he settled in his old home at Limerick, and again took up the uncongenial duties connected with the family business.

But soon a new prospect opened out. The retirement of William Allman left vacant the Chair of Botany in Dublin University. Harvey had little hesitation in applying for the post, to which, he points out to a friend, "a moderate salary and comfortable College-rooms are attached. It is an old bachelor place," he writes, "and would in many ways suit me very well. The only thing on the face of it disagreeable is the *lecturing*, but I don't think I should mind that much, as it is lawful to have the subjects for the class written down." Harvey's candidature was viewed favourably by the University authorities, but a difficulty arose, inasmuch as the School of Physic Act prescribed that the Professor of Botany should hold a medical

[1] *Flora Capensis*, Vol. I. p. 8*.

degree, or the licence of the College of Physicians. To render him eligible, the degree of M.D. was at once conferred on Harvey *honoris causa*, but after a good deal of discussion this solution of the question was held to be inadmissible, and George James Allman was appointed to the vacant chair. Harvey, however, obtained the smaller appointment of Keeper of the University Herbarium, which had fallen vacant at about the same time owing to the death of Dr Thomas Coulter, the botanical explorer of Central Mexico and California.

Harvey now at last found himself in a congenial post, with a fair amount of leisure, and facilities for scientific work. He presented his herbarium of over 10,000 species to the University, which already possessed Coulter's extensive American collections. "I am as busy as a bee these times," he writes. "I rise at 5 a.m. or before it, and work till breakfast-time (half-past eight) at the 'Antarctic Algae[1].' Directly after breakfast I start for the College, and do not leave it till five o'clock in the evening. Again at plants till dusk. I am writing on the 'Antarctic Algae,' and arranging the Herbarium, and have been working at Coulter's Mexican and Californian plants." College vacations were now usually spent at Kew, staying with his best friend Sir William Hooker, and working hard in the Herbarium. On the way home from the first of these vacations, he went to Torquay, to spend some time with his old correspondent, Mrs Griffiths. They went out boating, he and the good lady of seventy-six years ; and together they visited the only British habitat of *Gigartina Teedii*, six miles away, and gathered that coveted sea-weed in the spot where Mrs Griffiths had discovered it in 1811, the year in which Harvey was born.

Another very rare alga which he received about this time, to his great delight, was *Thuretia quercifolia* from Australia, one of the most remarkable of sea-weeds, bearing oakleaf-shaped red fronds, formed of a beautiful lace-like double network with regular hexagonal openings, which he was himself destined to collect in quantity some years later at Port Phillip, and to figure in his *Phycologia Australica*[2].

[1] The algae of Beechey's Voyage.   [2] Vol. I., plate XL.

The circumstances under which this plant was found must have made Harvey's mouth water.

"My specimen," he writes, "was picked up by a lady who accidentally landed for a few hours in a little harbour, into which the ship put during a gale, and she describes the shore as covered with the most wonderful profusion of plants and animals. She got all the pocket handkerchiefs of the party and filled them with what came first to hand, and in this hasty way picked up sixty different kinds of sponges, forty of which are new species, and several Algae, among which was the above described beauty. Her husband (a captain) is going out again, and promises to gather all he can meet with. Don't I hope he may have a run in again in a squall!"

Harvey now commenced the publication of the first of his larger works on seaweeds—the classical *Phycologia Britannica*, a series of 360 coloured quarto plates, drawn on stone by his own hand, representing all the species then known to inhabit the British Isles, and accompanied by suitable letterpress: the whole taking five years to complete. This work represented an immense advance in the knowledge of British sea-weeds, and, by the beauty and excellence of its plates, did much to popularize the study of these interesting plants.

In the following year he began his *Nereis Australis, or Algae of the Southern Ocean.* This was the first fruits of a comprehensive scheme of publication, which in its entirety was to "form a compendious picture of the vegetation of the ocean," the *Nereis Australis* being followed by a similar *Nereis Tropica* and *Nereis Borealis*; but only a section of the scheme was carried out, and publication stopped with the issue of 120 pages of letterpress and fifty coloured plates, drawn as usual by Harvey himself. In 1849 he issued *The Sea-side Book*, a popular account of the natural history of the sea-shore, which ran through several editions.

About this time he secured an additional appointment which, while it added to his professional duties, also increased his opportunities for research. The Royal Dublin Society, founded in 1731 for the improvement of husbandry, manufactures, and other useful arts and sciences, and aided by considerable government funds, had long since embarked on comprehensive schemes for the development of both science and art. To its activity is due the foundation and building up of many of the leading educational institutions in Dublin—the National Museum, the National Library, the Botanic Gardens at Glasnevin, the Metropolitan School of Art. The Society had established also professorships of zoology, botany, natural philosophy, chemistry, and so on. In 1848 the professorship of botany became vacant by the death of Dr Samuel Litton, and Harvey applied for the post. These appointments were made by the vote of the members at large, and strongly against his inclination, he had to enter on a personal canvass, of some experiences of which he gives a half humourous, half pathetic account in a letter to N. B. Ward, of "Wardian case" fame, who throughout life was one of his most regular correspondents. The issue was satisfactory, Harvey being elected by a three-fourths majority. This appointment placed him in control of the Glasnevin Botanic Gardens, of which Dr David Moore, so well known by his work on the Irish flora, was curator. It made him responsible besides for the delivery annually of courses of botanical lectures in Dublin, and also, at intervals, in selected towns in various parts of Ireland.

In the spring of 1849 Harvey accepted an invitation from the Smithsonian Institution and Harvard University to deliver twelve lectures on botany at the Lowell Institute at Boston, and others at Washington. The subject he chose for the Boston course was a comprehensive survey of the plant-world, from the point of view of the "progressive organization of the vegetable entity." The cryptogams had a place of honour, four lectures being devoted to Algae : it is interesting to note that the Fungi, which he designates "the most aristocratic of Crypts—*fruges consumere nati*," he placed immediately below the Flowering Plants, for reasons which, no doubt, he gave in his discourses. He sailed from Liverpool in July. Ocean traffic had been revolutionized since his last voyage from the Cape; instead of a dawdling sailing-ship, a steamer transported him in ten days to Nova Scotia; and with some of the old excitement with which he had started on his first climb up Table Mountain, he rambled away into the dark spruce woods, through the rich undergrowth of Kalmias, Ledums and Andromedas, with Sarracenias and Orchids rising from among the Sphagnums in the damper spots. He dredged and shore-collected also, but the seaweed flora was not rich. Thence he passed to New York, which he describes as like twenty Birkenheads and a dozen Liverpools, with slices from London and Paris, all huddled together, and painted bright red, with green windows. He visited Niagara and Quebec, and then travelled to Boston, where he was welcomed by Asa Gray, who was his host during his stay.

The lectures were well attended, and Harvey seems to have been satisfied with them and with the reception which they received; a popular lecture on seaweeds at the Franklin Institute at Providence was largely attended. These discourses, and the introductions and conversations that ensued, had more than a passing interest, as recruits were enrolled for alga-collecting, who subsequently supplied valuable material for his work on North American seaweeds. He saw all that was best of scientific society in Boston and New York, and met many of the great men of that generation—Agassiz, Bailey, Dana, Longfellow, Leidy, Pickering, Prescott, Silliman, Daniel Webster, Oliver Wendell Holmes. Having fulfilled his engagements and revisited the family of his late brother Jacob in New York, he turned his face southward in January for a collecting tour along the Atlantic sea-board.

After brief stays at Wilmington and Charleston, where he did a little botanizing, and sent to Kew a box-full of Dionaea, he arrived by boat at Key West one Sunday midnight in pouring rain, to spend the remaining hours of darkness in wandering about seeking a lodging. But by morning his fortunes had mended, and he spent a busy and pleasant month there, collecting by day, dodging mosquitos by night, and living mainly upon turtle and roast turkey, more ordinary foods being scarce. He made large collections of Algae, almost every day bringing to him new and beautiful forms. He had hoped to have the company of Prof. Bailey on this trip, but illness prevented this, and he had to carry out his work alone.

Màrch saw him back in Charleston, where he attended the annual meeting of the American Association for the Advancement of Science. Then to Washington, where he delivered four lectures at the Smithsonian Institution. At Charleston he again met Agassiz, and once more records the profound impression which the American zoologist produced upon him. " His fine thought," he writes, " of reforming the classifications of animals by a more intimate study of their young in the various stages from embryonic life to full development, grows apace; and if he lives to bring out his conception of a system based upon this, it will not only crown his memory for ever, but be the greatest step of the present age in zoological science...He is certainly a man of extraordinary genius, great energy, and with the most rapid inductive powers I have ever known. I could not help saying to myself, as I sat and listened, Well, it is pleasant to be hearing all this, as it is uttered, and for the first time. If one lives to be an old man, one will have to say, ' I remember to have heard Agassiz say so and so,' and then every one will listen, just as we should do to a person who had conversed with Linnaeus or Cuvier." We must remember that this appreciation of Agassiz's ideas was written nine years before the publication of Darwin's *Origin of Species*, and at a time when American men of science were much interested in a controversy as to whether mankind are all descended from Adam and Eve, or from several separate creations in different parts of the world. One of his last letters written on American soil contains a note on another subject, significant in the light of subsequent events. " I have been twice at sittings of the Senate, and have heard a good sensible speech on the Union question, which is now agitating folk here...The bone of contention is Slavery."

The spring of 1850 saw him once again settled in Dublin, with a great accumulation of work on hand. Part of the summer was spent in collecting Algae on the coast of Antrim; and he met again his friends Asa Gray and his wife, who were visiting Europe. Another acquaintance made at this time, which ripened into a warm friendship, was the result of the finding by Mrs Alfred Gatty, well-known as a writer of fiction, of the *Chrysymenia orcadensis* of Harvey at Filey, in fruit for the first time—the examination of which convinced Harvey that the Orkney plant was only a variety of *Chrysymenia rosea* (*Lomentaria rosea* Thuret). Mrs Gatty became a useful ally in the collecting of seaweeds, and a valued friend; Harvey's influence is seen in her *British Seaweeds*, published in 1863.

The year 1851 saw the completion of the *Phycologia Britannica*, and he at once set to work on his *Nereis Boreali-Americana*, published in three parts in the *Smithsonian Contributions to Knowledge*—a work of 550 quarto pages containing an account of all the known species of North American Algae, and 50 coloured plates, lithographed as usual with his own hand—a fine piece of work, and one which has not yet been superseded. This was a time of strenuous labour, for already he was planning a still more extended foreign tour; but he found time in the autumn of 1852 for a trip to Switzerland with Sir William Hooker and other friends.

In August, 1853, Harvey set out on the most extended scientific expedition of his life. So far his collecting had been done in Europe, South Africa, and North America. Now he was to visit the Indian Ocean and Australasia, and to investigate their seaweed flora, as yet but little known.

A short stay was made in Egypt, and a sea-shore ramble at Aden yielded *Padina pavonia* and a few other seaweeds, but otherwise he made no stop till Ceylon was reached. There he travelled a good deal, but seaweed collecting was not so successful as he had hoped. Some of the places explored proved unproductive, and the prevalence of the monsoon rendered collecting difficult or impossible. But the last three weeks, spent at Belligam Bay and Point de Galle, yielded excellent results, and he proceeded to Singapore en route for Albany, with a collection of about 5000 specimens of Algae.

The first work in Australia was done in the extreme south-west. Here he gathered seaweeds assiduously in King George's Sound, but the ground proved rather poor, though one welcome storm brought him a rich harvest, of which he preserved 700 specimens in one day. He moved on to Cape Riche, to the eastward, travelling through the bush on foot, and thus making intimate acquaintance with the interesting vegetation as well as the fauna of the district traversed. Cape Riche proved poor also, and he went northward to Perth, where he met James Drummond, the pioneer of West Australian botany, formerly of the Botanic Garden at Cork, and the discoverer of *Spiranthes Romanzoffiana* in the British Islands. At Perth he struck good ground. "This place is an excellent locality for Algae," he writes, "I am daily finding fresh ones, and have the prospect of a good harvest of novelty and interest...The days are too short for my work. My best collections are made at Garden Island, nine miles distant.. I have been twice landed for a two hours' walk, and on both occasions collected so much that it took three days to lay them on paper." Rottnest Island also proved highly productive, and he gives a very attractive picture of the great rock-pools on the limestone reefs, filled with brilliant seaweeds, many of them undescribed. Here he lived in the deserted convict establishment, and amassed a large and valuable collection.

Thence he went to Melbourne, where he collected at several points about Port Phillip, notably on Phillip Island; after which he sailed for Tasmania, where at Georgetown he had a month's successful work with the Rev. J. Fereday, himself an enthusiastic student of botany, seaweeds included. Passing through Hobart, he obtained permission to visit Port Arthur, at that time a great convict station, for which he sailed on March 1, 1855, passing the grand basaltic headlands of Cape Raoul and Cape Pillar. At Port Arthur amid exquisite natural surroundings marred by the presence of chained prisoners, armed warders, and sentry-lines of fierce dogs, he worked successfully, doing much shore-collecting, and dredging with the aid of a crew of convicts and armed guards. After a little rather unsuccessful collecting at Sydney and Newcastle he sailed for New Zealand, where he spent a few weeks at Auckland. While the terrestrial flora proved highly interesting to him, he found the shore poor in Algae; but he enlisted a useful recruit for collecting, in Mr Knight, Auditor-General, who undertook to collect and send him further material.

The 26th July, 1851, found him at Tonga Taboo, in the Friendly Islands, revelling in his first glimpse of nature in mid-Pacific. The fringing reef proved somewhat disappointing, for amid the multitudinous and many-coloured animal forms only a few green Algae were to be found. Harvey spent six months in the Pacific, visiting island after island according as the mission boats supplied a means of transport, collecting seaweeds and a good many marine animals. At that time social conditions in the South Seas were very different from what they are now. The adjoining Fiji Island group, for instance, was still in a savage state: the captain of the mission vessel told Harvey how, only four years before, he had seen one hundred human bodies laid out for a great feast, and cannibalism was still a habitual practice there; but the Friendly Islands, though but recently in a similar condition, seem already to have deserved their name, and Harvey's experiences of the natives, with whom he was much in contact, appear to have been of the pleasantest description; in Fiji also, where several weeks were spent, the founding of a Christian mission (permitted only two years before after eighteen years' refusal) had already greatly altered local practices; devil-worship and cannibalism were rapidly dying out. Harvey, applying at the mission station for a responsible guide, was furnished with a man entitled " Koroe," which, it appeared, was an honourable title " something equivalent to a C.B. in England," and bestowed only on a person who had committed at least five murders. Harvey returned to Sydney, and thence to Europe by Valparaiso and Panama, having a severe bout of fever on the way. He reached home in October, 1856, after an absence of over three years.

Here an important change of life awaited him. G. J. Allman succeeded to the Natural History chair in Edinburgh, rendered vacant by the death of Edward Forbes, and Harvey was elected to the chair in Trinity College, Dublin, the difficulties which led to his rejection twelve years earlier being not raised on this occasion, though the law remained the same. At the same time, the incorporation of the several Dublin Society professorships in the newly founded Museum of Irish Industry (now the Royal College of Science for Ireland), gave him additional work, as his post was converted into a Natural History and Economic chair. However, the considerable increase of lecturing and teaching thus brought upon him did not prevent his pushing on vigorously with the now large arrears of phycological work. His first action was to finish and publish the third and last section of the *Nereis Boreali-Americana* and then bring to a conclusion his enumeration of the seaweed flora of North America. This was accomplished in 1858, and in the same year he began the publication of the results of his work in Australia. The *Phycologia Australica*, which was issued in parts during the ensuing five years, ran to five volumes, each containing sixty coloured plates, and descriptions of all the species known from Australasian waters. In the year following the launching of this work, he commenced the publication of two important treatises on the phanerogamic flora of South Africa. In the first of these, the well-known *Flora Capensis*, he had the co-operation of Dr O. W. Sonder of Hamburg. This extensive work he did not live to complete; the third volume, which ran as far as the end of the Campanulaceae, being published the year before his death. The other work was his *Thesaurus Capensis*, a series of plates of rare or interesting South African plants, designed to supplement and illustrate the unillustrated *Flora*; of this he lived to issue only two volumes, each containing one hundred plates.

Harvey's home life, which for several years had been very lonely, was transformed in 1861, when, at the age of fifty, he was married to Miss Phelps of Limerick, whom he had long known. But almost immediately afterwards the shadow of death appeared, haemorrhage from the lungs warning him that his newly found happiness might not endure. After a summer spent at his favourite Miltown Malbay, on the wild coast of Clare, he was able to resume his college duties and his work on *Flora Capensis*. Although he never fully recovered his health, he laboured diligently at the works he had in hand. He had a noble example of continued devotion to science in his old friend Sir William Hooker, whom he again visited, on returning from a tour on the Continent, in the autumn of 1863, to find him, in his seventy-ninth year, finishing off the last volume of his *Species Filicum*, and "already beginning to nibble at another book." This was a further work on ferns, the *Synopsis Filicum*, on which Hooker was busily engaged until within a few days

of his death in the summer of 1865; it was completed by
J. G. Baker and published three years later. During the winter
of 1865, Harvey himself became seriously ill, and, an immediate
change to a mild climate being recommended, he and his wife
went to stay at Torquay with Lady Hooker, and there he died
on 15th May, 1866.

Harvey was only fifty-five years of age when he died, but
he had won for himself a foremost place among systematic
botanists. Life, as Lubbock has said, is measured by thought
and action, not by time; and according to this standard, Harvey's
life-cup was already full and running over. He had used to the
utmost the gifts which he possessed. The capital with which he
entered on his career comprised a critical eye, a deft hand, and
that scientific enthusiasm without which no botanist ever travels
far. On the other side of the account, he had two serious
deterrants, a rather delicate body, and a complete absence of
scientific training. "*Apropos* of dissection," he writes to Hooker
in his younger days, "I am a miserable manipulator, and should
be very grateful for a few lessons." From the beginning he had
a shrewd perception of what lay within his reach, and what was
beyond it. "The extent to which I mean to go in botany," he
wrote at twenty-one years, "is to know British plants of all
kinds as well as possible; to know Algae of all countries
specially well; to collect all foreign Cryptogamia that may fall
in my way, and to know them *moderately* well...My reason for
choosing the *Algae* is pure compassion; they being sadly
neglected by the present generation, though at a former time
they were in high favour."

In the letters written even in boyhood we see foreshadowed
the direction and extent of his future researches. "Exactly
what he determined in youth to accomplish," says Dr John
Todhunter in his Preface to Harvey's *Memoir*, "he accomplished;
the work which he took upon himself to do he did, honestly
and thoroughly; the fame which he desired to achieve, he
achieved." He saw that his strength lay in discrimination,
description, and illustration, and to these—the necessary census
task which forms the groundwork on which great theories may
be built up—he confined himself.

The latter years of his life fell within that stimulating period which followed the publication of Darwin's *Origin of Species*. But in the battle of giants which ensued he took no part. His attitude, indeed, was rather that of an amused spectator; and in the letters which are available, his references to the great controversy of the day, and allied topics, are mostly in a playful vein. "I do not know how cats purr," he writes to his friend Mrs Gatty, "and am glad you asked...Have you never felt a something stop your own windpipe when pleased or grieved, when suddenly affected either way? 'Tis the first gurgle of a purr; you were a cat once, away in the ages, and this is a part of the remains." Almost his only contribution to the literature of natural selection was a "serio-comic squib," which was read before the Dublin University Zoological and Botanical Association on 17 February, 1860 and subsequently printed for private circulation, entitled "A Guess as to the Probable Origin of the Human Animal considered by the light of Mr Darwin's Theory of Natural Selection, and in opposition to Lamarck's notion of a Monkey Parentage." Darwin thought this production a little unworthy of the author. "I am not sorry for a natural opportunity of writing to Harvey," he says, "just to show that I was not piqued at his turning me and my book into ridicule, not that I think it was a proceeding that I deserved, or worthy of him[1]."

Similarly, Harvey rejoices over Charles Kingsley's *Water Babies*, and especially over the sly fun which is poked at Darwinism, and also at certain types of men of science.

Only once did he enter the lists with a serious criticism, when, in the *Gardeners' Chronicle*[2], he cites the case of a monstrous Begonia in objection to Darwin's views. Harvey, indeed, did not like the new theory. "I am fully disposed to admit natural selection as a *vera causa* of much change," he writes, "but not as the *vera causa* of species." Further than this he could not go, though much impressed with the arguments drawn from geographical distribution. "I heartily wish we were nearer in accord," writes Darwin at the end of a long letter to Harvey,

---

[1] Darwin's *Life and Letters*, Vol. II. p. 314.

[2] For 1860, pp. 145–146.

"but we must remain content to be as wide asunder as the poles, but without, thank God, any malice or other ill-feeling[1]."

Thus it will be seen that Harvey took but little part in influencing the thought of his time; the materials for his work were gathered not from his own creative brain, nor from the thoughts of other men, but direct from Nature's storehouses; his study was the far-stretching shore, his companions

> "The toiling surges,
> Laden with sea-weed from the rocks,"

his duty the describing with pen and pencil the harvest of the sea. In his works, he rises above mere technical description of the species with which he is dealing. His mind is filled with the beauty and wonder of plants; and he strives to impress the reader with the deep interest of the study of botany. He endeavours always to popularize his favourite pursuit by means of pleasant general introductions, and to promote a better knowledge of seaweeds or of flowering plants by appealing to his readers to collect, and by giving instructions for the gathering and preserving of specimens.

He derived a peculiar satisfaction from the thought that, at his post at Trinity College, Dublin, he was building up a great permanent collection that would be useful to future generations of botanists. "Here," he writes, "I sit like a turnspit roasting the meat, and when I am gone I suppose another dog will be put in my place. The Herbarium will not be broken up. I am content, for I seem to be working for some little purpose. I should just like to leave it in better order—to get through the arrears—and to return borrowed specimens." It was the same thought that prompted him to the publication of the great descriptive works which his rapidity and skill with pen and pencil enabled him to complete despite frequent intervals of illness. He devoted himself to his task with intense application. "Twenty minutes," he writes from South Africa in the middle of the stifling summer, "is my fair allowance for a drawing, with all its microscopical analysis."

From his letters, and from the reminiscences of persons who

---

[1] *More Letters of Charles Darwin*, Vol. I. p. 166.

remember him, one gathers that Harvey was a very lovable sort of man. Shy and retiring, and diffident as to his own powers, with a deeply affectionate nature, he was equally prone to singing the praises of his friends, and to disparaging himself. "If I lean to glorify any one," he writes to William Thompson of Belfast, "it is Mrs Griffiths, to whom I owe much of the little acquaintance I have with the variations to which these plants [the seaweeds] are subject, and who is always ready to supply me with fruits of plants which every one else finds barren. She is worth ten thousand other collectors." Writing of Harveya, a genus of South African Scrophularineae which Hooker had just named in his honour, he comments, "'Tis *apropos* to give me a genus of Parasites, as I am one of those weak characters that draw their pleasures from others, and their support and sustenance too, seeing I quickly pine, if I have not some one to torment." He in his turn loved to commemorate his friends, or others in whom he felt an interest, by naming after them new genera of plants—Apjohnia, Areschougia, Ballia, Backhousia, Bellotia, Bowerbankia, Drummondita, Curdiea, Greyia, Mackaya, and many others. The names of some of his favourite authors are similarly enshrined, as Crabbea, Evelyna. Indeed, when at Niagara he saw an inscription to a young lady who fell over the cliff when gathering flowers—

> Miss Ruggs at the age of twenty-three
> Was launched into eternity,

he comments "Poor thing! I must call a plant after her— *Ruggia* would sound well." He had indeed a love of all living things. Writing to Mrs Gray on the death of her favourite dog, he tells how he felt so ashamed of being so deeply moved when in South Africa by the death of his pet ostrich, that he foreswore any similar entanglement, and kept his vow ever since. Of serious griefs he had many; the death of several beloved brothers and sisters who predeceased him, would have been well nigh intolerable to him but for the profound religious feeling which sustained and helped him throughout life, and which robbed death of all its terrors.

I cannot do better than conclude with some words in which

Asa Gray summed up Harvey's work and character shortly after his decease[1]: "He was a keen observer and a capital describer. He investigated accurately, worked easily and readily with microscope, pencil, and pen, wrote perspicuously, and where the subject permitted, with captivating grace; affording, in his lighter productions, mere glimpses of the warm and poetical imagination, delicate humour, refined feeling, and sincere goodness which were charmingly revealed in intimate intercourse and correspondence, and which won the admiration and the love of all who knew him well. Handsome in person, gentle and fascinating in manners, genial and warm-hearted but of very retiring disposition, simple in his tastes and unaffectedly devout, it is not surprising that he attracted friends wherever he went, so that his death will be sensibly felt on every continent and in the islands of the sea."

[1] *American Journal of Science and Arts*, Vol. XLII. p. 277.

Plate XVIII

MILES JOSEPH BERKELEY

MILES JOSEPH BERKELEY
1803—1889

By GEORGE MASSEE

Narrative—early interest in Natural History—Zoological publications—
Algae—Fungi—character and magnitude of Berkeley's work in systematic Mycology—exotic fungi—co-operation with Broome—morphology
of Basidiomycetes—*Introduction to Cryptogamic Botany*—pioneer work
in plant pathology—the potato disease—personal characteristics.

MILES JOSEPH BERKELEY was born at Biggin Hall, near
Oundle, Northamptonshire, on the 1st April, 1803. He was the
second son of Charles Berkeley, whose wife was a sister of
P. G. Munn, the well-known water-colour artist. His family
belonged to the Spetchley branch of the Berkeleys, and had
been resident for several generations in Northamptonshire.
Berkeley received his preliminary education at the Oundle
Grammar School and afterwards at Rugby, entered Christ's
College, Cambridge, in 1821, and graduated as 5th Senior
Optime in 1825. He was ordained in 1826, and his first clerical
duty was the curacy of St John's, Margate. In 1833 he became
Perpetual Curate of Apethorpe and Wood Newton, Northamptonshire, and resided at the neighbouring village of King's
Cliffe, a name familiar to every mycologist as being the habitat
of numerous species of fungi, first recorded as members of the
British Flora. In 1868 he was appointed Vicar of Sibbertoft,
near Market Harborough, where he died on the 30th July, 1889,
at the age of 86 years.

As a boy Berkeley was much devoted to the study of nature,
paying special attention to the structure and habits of animals;
he also at an early age made a somewhat extensive conchological

collection. This tendency was to some extent fostered at Rugby, but the influence exercised by Professor Henslow during Berkeley's time at Cambridge, and the opportunities of studying the progress of research made in the various branches of Natural History, were the chief factors that determined Berkeley to enter seriously on the study of what at the time was styled Natural History.

His first published paper was "On new species of *Modiola* and *Serpula*" (*Zoological Journal*, 1828). It was followed by "On the internal structure of *Helicolimax Lamarckii*"; "On *Dentalium subulatum*"; "On the animals of *Voluta* and *Assiminia*" (idem 1832–34); and "On British *Serpulae*" and "*Dreissenia polymorpha*" (*Magazine of Natural History*, 1834–36).

A series of 'beautifully executed coloured drawings and dissections, illustrating Berkeley's zoological studies, may be seen at the Herbarium, Kew. Although all Berkeley's publications up to this time dealt with zoological subjects, yet the study of Botany had been by no means neglected, and about this time having made the acquaintance of Dr Harvey of Dublin, Dr Greville of Edinburgh, the author of *Scottish Cryptogamic Flora*, and of Captain Carmichael of Appin, N.B., a trio of the most celebrated cryptogamists of the age, Berkeley forsook the serious study of zoological subjects, and devoted the whole of his leisure time to the lower forms of plant life. Living at Margate, the marine algae naturally attracted Berkeley's attention, and in 1833 he published his *Gleanings of British Algae*, consisting of a series of detailed investigations on the structure of the minute and obscure forms of marine and freshwater species. This work, illustrated by twenty coloured plates, was originally intended to be included in the supplement to Dr Greville's *Scottish Cryptogamic Flora*, but in consequence of the discontinuance of that most excellent work, was issued as an independent booklet.

From the first Berkeley was deeply interested in the fungi, and practically all his subsequent work was devoted to this group of plants, and although well versed in general Cryptogamic Botany, it was in the field of Mycology that his laurels were

won. A review of the work done can be most conveniently
discussed under three separate headings—Systematic Mycology,
Morphology and Literature, and Plant Pathology, respectively.

Systematic Mycology.

Under the title *British Fungi*, four fascicles of dried and
well-prepared specimens, numbering in all 350 species, were
issued between 1836 and 1843. In those days exsiccatæ were
not issued from a commercial standpoint, as is too frequently
the case at the present day, but represented the outcome
of careful investigation on the part of the author, hence
Berkeley's exsiccatæ are at a premium at the present day.

In 1828 Berkeley first corresponded with Sir W. J. Hooker
on matters dealing with cryptogams, and in one of his early
letters stated that he had devoted much time to the study of
fungi, more especially to the extensive genus *Agaricus*, which at
that period included all the gill-bearing fungi. At this time,
Sir William was engaged in preparing the volumes dealing with
cryptogams, as supplementary to *The English Flora* of Sir
James Edward Smith, and approached Berkeley on the subject
of undertaking the section dealing with Agarics, in the volume
devoted to the fungi. Berkeley agreed to this arrangement,
and was finally induced to describe the whole of the fungi.
A footnote at the commencement of the volume by Sir W. J.
Hooker is as follows:

"When the printing of the species of this, the 2nd Part of
the Class *Cryptogamia*, was commenced, I thought myself highly
fortunate to have obtained the assistance of my valued friend,
the *Rev. M. J. Berkeley*, in preparing the first Tribe, *Pileati*.
I have now to express my cordial acknowledgements (in which
I am satisfied I shall be joined by every Botanist in the country)
to that gentleman for having kindly undertaken to prepare the
whole of this vast family for the press: and it is certain that the
task could not have fallen into better hands."

The volume contains detailed descriptions of all British
fungi known at the time, amounting to 1360 species, included in
155 genera, the great majority of which had been studied by

the author in a living condition, and also compared with specimens contained in various exsiccatæ and with the very extensive collection owned by Sir W. J. Hooker. The appearance of this book at once placed Berkeley in the front rank of Mycologists, and it was universally admitted as the most complete Mycologic Flora of any country extant; and furthermore, so far as accurate information, and a true sense of the conception of species are concerned, the same statement holds good at the present day. At this date our knowledge of extra-European fungi was almost nil, with the exception of a few woody cosmopolitan species collected by various travellers, more as matters of curiosity than for the advancement of our knowledge of the fungus-flora of the world.

Opportunity alone was required by Berkeley, and such opportunity was readily afforded by Sir W. J. Hooker, who placed unreservedly in Berkeley's hands the various collections of exotic fungi received at Kew from time to time. This practice was continued by the two succeeding Directors at Kew, Sir Joseph Dalton Hooker and Sir William Thiselton-Dyer. Such unrivalled opportunities were utilised to the fullest extent by Berkeley, who soon manifested by his treatment of the material placed in his hands a thorough grasp of the subject, and for nearly half a century practically all collections of exotic fungi passed through Berkeley's hands. During this period 6000 new species were described, and in numerous instances illustrated, including many new genera from all parts of the world, arctic, antarctic, tropical and temperate. Botanists were now enabled, for the first time, to grasp the true significance of the fungus-flora of the world, which numerically ranks next to Phanerogams, and which was shown to exercise an influence on life on the globe in general, not realised before Berkeley's time. The better known European genera of fungi, many of which appeared to be sharply defined, and by some mycologists considered to be of ordinal importance, could now be estimated at their true value and relegated to their true position in the scheme of classification rendered possible by a good knowledge of the range of structure presented by the fungi of the world at large. As regards geographical distribution, Berkeley repeatedly emphasized the fact that the fungi are more cosmopolitan than any other known group of plants, and that their abundance at any place during a given period was almost entirely dependent on conditions favouring the development of the higher forms of plant life, fungi only following in the wake of such, and never posing as pioneers, on account of the nature of their food. Amongst the numerous novel types of extra-European fungi described by Berkeley, it is somewhat difficult to indicate briefly even a few of the most striking forms. Perhaps his genus *Broomeia* stands out pre-eminent. It belongs to the puffball group of fungi, and is unique in that family—the Gasteromycetaceae—in having numerous individuals springing from, and imbedded in a common sterile base or stroma. It is a native of the Cape of Good Hope. The following is Berkeley's dedication of this genus to his friend and co-worker, C. E. Broome, M.A., of Bath. "Nomen dedi in honorem amicissimi, C. E. Broome, armigeri, Tuberacearum Anglicarum accuratissimi indagatoris, cujus pene solius laboribus extant hodie viginti species indigenae fungorum hypogaeorum." *Broomeia congregata* Berk., is described and figured in Hooker's *London Journal of Botany*, 1844. Certain club-shaped fungi parasitic on caterpillars, belonging to the genus *Cordyceps*, occurring on buried caterpillars in New Zealand, are the giants of their tribe, measuring up to eighteen inches in length. Finally, Berkeley first introduced to our notice many of those quaint fungi belonging to the group including our well known "stinkhorn"—*Phallus impudicus* L.—and cleared up many points in their structure previously unknown. Fries, the most distinguished mycologist of his time, writes as follows in his Preface to *Hymenomycetes Europaei*; "Desideratissima vero Synopsis Hymenomycetum extra-europearum, qualem solus praestere valebit Rev. Berkeley."

Notwithstanding Berkeley's researches on exotic fungi, a task in itself too comprehensive for most men to grapple with, he continued to study the British fungi, and, mostly in collaboration with his friend, Mr C. E. Broome, published a long series of articles in the *Annals and Magazine of Natural History*, from 1837 down to the year 1883. In these articles 2027 species of fungi are enumerated, mostly new, or species new to Britain, and consist mainly of critical notes on the morphology and affinities of the fungi under consideration, and will compel the attention of mycologists for all time.

From the above brief account it may perhaps be concluded that Berkeley was essentially a systematist and founder of new species. Owing to the vast amount of material that passed through his hands, he was so perforce, but his leaning was always rather towards the biological and morphological side of the subject.

Morphology and Literature.

The first important paper dealing with the morphology of the hymenial structure in Fungi, is entitled, "On the Fructification of the Pileate and Clavate Tribes of the Hymenomycetous Fungi," *Annals of Nat. Hist.*, 1838. Here is clearly demonstrated for the first time, the universal occurrence of basidia bearing spores at their summit, throughout the entire group of fungi known to-day as the Hymenomycetes, including Agaricaceae, Thelephoraceae, Clavariaceae, etc. This important discovery rendered possible the basis of a classification on morphological grounds, which holds good at the present day. A careful study of the text and illustrations demonstrates the fact that Berkeley was perfectly well acquainted with all the essential details of the hymenium, many of which have been repeatedly rediscovered and described under new names, in ignorance of the fact that such structures had previously been equally well described.

Berkeley continued his investigations on the structure of the hymenium, and his next paper, entitled "Sur la fructification des genres *Lycoperdon*, *Phallus* et de quelques autres genres voisins," in *Annal. Sci. Nat.* Ser. 2, vol. XII. (1839), demonstrated the universal presence of basidia bearing spores at their summit in the family now known as the Gasteromycetes. This research on the part of Berkeley led to the universal adoption of the two primary divisions of the Fungi ; Basidiomycetes, having the spores borne at the apex of a basidium; and Ascomycetes, having the spores produced within specialised sacs, or asci.

In 1857 the *Introduction to Cryptogamic Botany* appeared, which remained for many years the standard work on the subject. This was followed in 1860 by *Outlines of British Mycology*, a book profusely illustrated with coloured plates, and intended more especially for the beginner in the study of Mycology.

Just over 400 separate papers dealing with fungi are listed under Berkeley's name alone, in addition to numerous others, where he worked in collaboration with C. E. Broome, Dr M. C. Cooke, Rev. M. A. Curtis, and others.

Plant Pathology.

At the present day Berkeley is best known as a systematist, which of itself alone is sufficient to retain his name for all time in the front rank of mycologists, but when the history of Plant Pathology is elaborated, Berkeley's name will undoubtedly stand out more prominently than that of any other individual. In fact, it is not saying too much to pronounce Berkeley as the originator and founder of Plant Pathology. He was not the first to investigate plant diseases caused by fungi, but he was undoubtedly the first to recognise the significance of the subject, and its great importance from an economic standpoint. His investigation of the potato murrain, written in 1846, cleared the air of all kinds of wild theories as to its origin, and showed it to be undoubtedly caused by the fungus now known as *Phytophthora infestans*, whose life-history he carefully worked out. Then followed a similar investigation of the vine-mildew, and a series of researches on diseases of plants published in the *Gardeners' Chronicle* dating from 1854 to 1880. It was in these numerous communications that the science of Plant Pathology was firmly established and propounded. The article "On the Diseases of Plants" was contributed to the *Cyclopaedia of Agriculture* by Berkeley.

In 1879 he unconditionally presented his mycological herbarium to Kew. This collection contained 10,000 species, of which 5000 were types of Berkeley's own species, in addition to numerous co-types from Montagne, Schweinitz, Fries, Cooke and other contemporaneous mycologists. Hence Kew is, and must for ever remain, the Mecca of mycologists from all parts of the world.

Berkeley was a man of great refinement, and an excellent classical scholar. His tall commanding figure and grand head with flowing white hair, as I knew him late in life, could not fail to arrest attention. Unobtrusive and by no means ambitious, and too enthusiastic to be self-seeking, Berkeley was tardily promoted to the Honorary Fellowship of his College, and elected a Fellow of the Royal Society at the age of 76. In 1876 a Civil List Pension of £100 per annum was awarded, for his services to botany with especial reference to his investigations on the diseases of plants.

Plate XIX

JOSEPH HENRY GILBERT

SIR JOSEPH HENRY GILBERT

1817—1901

By W. B. BOTTOMLEY

Early training in Chemistry—his meeting with Lawes—official distinctions—
the Lawes-Gilbert combination—the Rothamsted Reports—Liebig's
'mineral theory'—the relation to nitrogen—Leguminous plants—
Hellriegel and others—confirmation of their results—nitrification—
feeding of stock.

JOSEPH HENRY GILBERT was born at Hull on August 1,
1817. He was a son of the manse being the second son of the
Rev. Joseph Gilbert, a Congregational Minister. His mother was
one of the gifted daughters of the Rev. Isaac Taylor of Ongar,
and a well-known writer of hymns and songs for children.
Whilst at school young Gilbert had the misfortune to meet
with a gunshot accident which deprived him of the use of one
eye, a mishap which for a time threatened to mar his future
career, but his own inherent determination and the home-train-
ing of the manse enabled him to overcome the disadvantage of
defective eye-sight, and triumph over physical disability.

From school he went to Glasgow University and studied
chemistry under Professor Thomas Thomson, then to University College, London, where he attended the classes of Professor
Graham and others, and worked in the laboratory of Professor
Todd Thomson. Here it was, in Dr Thomson's laboratory,
that he first met Mr J. B. Lawes, with whom he was afterwards
so intimately associated. He then proceeded to Giessen for a
short time, studying under Liebig and taking his degree of
Doctor of Philosophy in 1840. Returning to London, he worked
at University College, acting as laboratory assistant to Professor

Thomson, and became a Fellow of the Chemical Society on May 18, 1841, when the Society was barely three months old. He then left London to take up calico printing and dyeing in the neighbourhood of Manchester, but returned south in 1843, at the invitation of Mr Lawes, to assist in the agricultural investigations at Rothamsted, Herts.

Mr John Lawes had begun experiments in 1837 on growing plants in pots with various manures. He discovered the fact that mineral phosphates when treated with sulphuric acid yielded a most effective manure. Taking out his patent for the production of superphosphates in 1842, Lawes soon found himself busy with the establishment of a successful business. Not wishing to give up the agricultural investigations which he had commenced in the fields of Rothamsted he decided to obtain scientific assistance, and remembering the young chemist he had met in Dr Thomson's laboratory, Gilbert was invited in June 1843 to superintend the Rothamsted experiments. Thus began that partnership in investigation which has yielded such a rich harvest of results, and an association with Rothamsted which lasted for fifty-eight years.

Gilbert was elected a Fellow of the Royal Society in 1860, and received a Royal Medal in 1867. He was President of the Chemical Section of the British Association in 1880, and President of the Chemical Society, 1882–3. In 1884 he was appointed Sibthorpian Professor of Rural Economy at Oxford, and held the chair until 1890. He was a member of various foreign academies and societies, and was the recipient of honorary degrees from several home universities, becoming LL.D. of Glasgow (1883), M.A. of Oxford (1884), LL.D. of Edinburgh (1890), and Sc.D. of Cambridge (1894). In 1893 on the occasion of the jubilee of the Rothamsted experiments he received the honour of knighthood.

The character and scope of Gilbert's life-work was well described by Prof. Dewar at a special meeting of the Chemical Society in 1898, when he said, "The work of Gilbert, as we know, was early differentiated into that most complex and mysterious study, the study of organic life. For the last fifty years he has devoted his attention to the physiology of plant

life in every phase of its development. With a skill that has
been unprecedented, he has recorded from year to year the
variations in the growth of every kind of nutritious plant. He
has examined into the meteorological conditions, the variations
of climate, of soil, and of mineral agents, of drainage, and of every
conceivable thing affecting the production and development of
plant growth. These memoirs are admitted throughout the world
to be unique in their importance. Wherever the chemist or the
physiologist, the statistician or the economist has to deal with
these problems, he must turn to the results of the Rothamsted
experiments in order to understand the position of the science
of our time. These results will be for ever memorable; they
are unique and characteristic of the indomitable perseverance
and energy of our venerated President, Sir Henry Gilbert."

The close association of Lawes and Gilbert in the Rothamsted experiments makes it almost impossible to separate the
work of the two men. The majority of the 132 papers issued
from Rothamsted between 1843 and 1901 appeared under the
joint names of Lawes and Gilbert, and it would be as difficult
as it is undesirable to attempt an analysis of this partnership.
It was essentially a partnership devoid of any jealousy, and
actuated by a feeling of mutual regard and esteem. There
never was a question as to the "predominant partner." The
two workers formed an unique combination, each supplying
some deficiency in the other. Lawes possessed the originating
mind and had a thorough knowledge of the facts and needs of
practical agriculture; Gilbert was the exact scientist, the man of
detail and method. Dr J. A. Voelcker, who speaks of Gilbert
as his life-long friend and teacher, says, "The partnership and
collaboration of 'Lawes and Gilbert' represented an excellent
embodiment of the motto 'Practice with Science.' Lawes was
essentially the practical agriculturist—quick to see and grasp
what the farmer wanted, and to become the interpreter to him.
He was the man to whom the practical farmer turned, the one
to write a brisk article on some subject of agricultural practice
or economy, to answer a practical question, or to solve some
knotty problem. Lawes was the more versatile of the two, the
more inclined to introduce changes in and modifications of the

original plan; and he has been known to say, jokingly, that if he
had been left to have his own way, he would have ploughed up
many of his experimental plots before they had yielded the full
results, which continuance on the old lines alone brought out.
Gilbert, on the other hand, was possessed of indomitable perseverance, combined with extreme patience and careful watching
of results.   His was the power of forecasting, as it were, what
might, in the end, lead to useful results.   With the determination to carry out an experiment to the very close he united
scrupulous accuracy and attention to detail.   Gilbert, it may be
said, was not so much the man for the farmer, but for the
scientist, and he it was who gave scientific expression to the
work at Rothamsted, and who established field experiments on
a scientific basis in this country."

To describe in detail Gilbert's work it would be necessary to
write an account of the Rothamsted experiments, a task beyond
our present limits seeing that the collected reports occupy nine
volumes.

The last published "Rothamsted Memoranda" gives a list of
132 papers.   They are divided into two series, one relating to
plants, the other to animals.

Series I. deals with "Reports of Field Experiments, Experiments on Vegetation, &c., published 1847—1900 inclusive," and
contains 101 papers.   These reports on plants are concerned
chiefly with the results obtained by growing some of the most
important crops of rotation separately, year after year, for many
years in succession, on the same land without manure, with
farm-yard manure, and with various chemical manures, the same
description of manure being, as a rule, applied year after year
on the same plot.

Amongst the numerous field experiments conducted on
these lines one of the most interesting is the field known as
Broadbalk field, in which wheat has been grown continuously for
over 60 years.   The results show that wheat can be grown for
many years in succession on ordinary arable land if suitable
manure be provided and the land be kept clean.   Even without
manure of any kind the average produce for 46 years—1852
to 1897—was nearly 13 bushels per acre, about the average yield

per acre of the wheat lands of the world. On this field it was found that mineral manures alone gave very little increase, whilst nitrogenous manures alone gave a much greater increase than mineral manures alone, but the mixture of the two gave much more than either alone. It is estimated that the reduction in yield, due to exhaustion, of the unmanured plot over 40 years —1852 to 1891—was, provided it had been uniform throughout, equivalent to a decline of one-sixth of a bushel per acre. It is related that a visitor from America, when being shown over the Broadbalk field, said to Sir John Lawes, "Americans have learnt more from this field than from any other agricultural experiment in the world."

Another set of field experiments of exceptional interest is that relating to the "Mixed Herbage of Permanent Grass Land." The land was divided into twenty plots. Two plots have received no manure from the commencement of the experiment, two have received a dressing of farm-yard manure each year, whilst the remainder have each received a different kind of artificial or chemical manure, the same kind being applied year after year on the same plot, except in a few special cases. Repeated analyses have shown how greatly both the botanical constitution and the chemical composition of the mixed herbage varied according to the kind of manure applied.

The results of these experiments were given under three headings—agricultural, botanical and chemical, and show in an exceptional manner the care of detail to which every investigation was subjected by Gilbert. Some people have thought that this minute attention to detail was carried to excess by Gilbert, and resulted in a bewildering multiplication of numerical statements and figures. One can, however, but admire his love of accuracy and absolute conscientiousness, and if his caution appeared at times to be carried to an extreme, the result has been to make "the Rothamsted experiments a standard for reference, and an example wherever agricultural research is attempted."

One of the most important results of the Rothamsted investigations has been the replacing of the "mineral theory" of Liebig by the "nitrogen theory" of Lawes and Gilbert. Liebig held the view that each crop requires certain mineral elements from the soil, and that crops will not flourish where the appropriate elements are lacking. Every soil contains some element in the minimum. Whatever element this minimum may be it determines the abundance and continuity of the crop. The only fertiliser which acts favourably is that which supplies a deficiency of one or more of the food elements in the soil. The atmosphere, according to Liebig, supplies in sufficient quantity both the carbon and nitrogen required by crops, and the function of manure is to supply the ash constituents of the soil. The exhaustion of soils is to be ascribed to their decreased content of mineral ingredients rather than to decrease in nitrogen.

When careful study of the composition of the atmosphere proved that the amount of ammonia brought down to the earth by rain scarcely exceeds a few pounds per acre annually, Liebig maintained that plants are capable of directly absorbing ammonia by means of their leaves. He pointed out that the beneficial effects of nitrogenous manures are most apparent in the case of cereal crops with a comparatively short vegetation period, and least apparent in the case of leafy crops with a long vegetation period. The long vegetation period of crops like clover allowed time for the utilisation of the ammonia of the air and no artificial supply was necessary. On the other hand, crops with a short vegetation period had a limited time for accumulating ammonia from the air, and responded readily to applications of nitrogenous manures.

Gilbert, early in his work at Rothamsted, noticed that the results of his field experiments were at variance with this "mineral theory," as it was called, of Liebig, and soon found himself involved in a controversy with the great German chemist which was not always free from bitterness. He found that the nitrogen compounds of the atmosphere were sufficient only for a very meagre vegetation. Cereals treated with ammonium salts and other nitrogenous manures showed a far greater increase of produce than when phosphates, potash or other ash constituents only were supplied. "As more nitrogen was assimilated a greater amount of the fixed bases were found in the ash, and he considered that the function of the fixed bases was to act as carriers of nitric acid. These bases—potash, soda, lime and magnesia, were not mutually replaceable, but the predominance of one or the other affected the produce. Luxuriance of growth was associated with the amount of nitrogen available and assimilated, and in the presence of this sufficiency of nitrogen the formation of carbohydrates depended on the amount of potash available." The possibility that the free nitrogen of the air might supply the nitrogenous needs of plants was disproved by growing plants in calcined soil and removing all traces of ammonia from the air before it was admitted into the glass case in which the plants were growing. Determinations were made of the nitrogen in the seed and soil at the beginning of the experiments, and in the plants and soil at their conclusion.

The work on the assimilation of nitrogen by plants extended over three years and was made the subject of a communication to the Royal Society in 1861. The paper, entitled, "The Sources of the Nitrogen of Vegetation ; with special reference to the question whether Plants assimilate free or combined Nitrogen," occupies 144 pages of the *Philosophical Transactions*, and is a brilliant example of the scrupulous accuracy and attention to detail which characterised all Gilbert's work. It is divided into two parts—I. "The General History and Statement of the question."—II. "The Experimental Results obtained at Rothamsted during the years 1857, 1858 and 1859." The authors state in the summary of conclusions that "in our experiments with graminaceous plants, grown both with and without a supply of combined nitrogen beyond that contained in the seed sown, in which there was great variation in the amount of combined nitrogen involved and a wide range in the conditions, character and amount of growth, we have in no case found any evidence of an assimilation of free or uncombined nitrogen.

"In our experiments with leguminous plants the growth was less satisfactory, and the range of conditions possibly favourable for the assimilation of free nitrogen was, therefore, more limited. But the results recorded with these plants, so far as they go, do not indicate any assimilation of free nitrogen. Since, however, in practice leguminous crops assimilate from some source so very much more nitrogen than graminaceous ones under ostensibly equal circumstances of supply of combined nitrogen, it is desirable that the evidence of further experiments with these plants under conditions of more healthy growth should be obtained."

As long as Gilbert's investigations were confined to non-leguminous plants and to leguminous plants grown in calcined soil the "nitrogen theory" was triumphant. When, however, leguminous plants were grown in uncalcined soil or in the open the results were uncertain, and in many cases the manures supplying ash constituents alone proved the most effective. The elucidation of these uncertain results has been a tedious problem, and has taken many years of patient investigation, but gradually the evidence accumulated which led to its solution.

Field and pot experiments in Germany, France, England and the United States in the late seventies and early eighties furnished abundant proof that under certain conditions leguminous plants do obtain nitrogen from the atmosphere, and gradually, from the work of Rautenberg, Frank and others, the idea was evolving that fungi or micro-organisms play some important part in the process.

Gilbert, however, would not listen to any such heresy, as he considered that the question of the assimilation of the free nitrogen of the air by plants had been finally settled by the experiments of 1857–60. It was therefore a most happy chance that Gilbert was present at the scientific congress in Berlin in 1886 when Hellriegel described his experiments on leguminous plants, showing that the formation of nodules on these plants was associated with the fixation of atmospheric nitrogen. In commenting subsequently on these experiments, Gilbert said, "It must be admitted that Hellriegel's results, taken together with those of Berthelot and others, do suggest the possibility that, although the higher plants may not possess the power of directly fixing the free nitrogen of the air, lower organisms, which abound within the soil, may have that power, and may thus bring free nitrogen into a state of combination within the soil in which it is available to the higher plants—at any rate to

members of the Papilionaceous family.   At the same time, it will
be granted that further confirmation is essential before such
a conclusion can be accepted as fully established."

This comment reveals the essential conservatism of Gilbert's
mind, but the true greatness of the man is seen when we
find him, at the age of seventy, repeating the experiments of
Hellriegel and Wilfarth, and himself supplying the confirmation
of their results which he considered essential.

The results of these experiments, contributed to the Royal
Society in 1887, 1889, and 1890, fully confirmed the theory that
leguminous plants are able to assimilate the free nitrogen of the
air by means of the micro-organisms contained in their root
nodules, and also explained the failure in the 1857-60 experiments to demonstrate nitrogen fixation by leguminous plants
owing to the use of calcined soil by which the inoculating
organisms present in the soil were destroyed.

Gilbert's investigations from 1871-75 showing that the
drainage waters from the experimental fields of Rothamsted
contained more nitrates as the amount of ammonium salts
applied to the soil increased, have been quoted by some writers
as being the basis of the modern theory of nitrification.   It must
be remembered that Gilbert was at first actively hostile to the
bacterial theory of nitrification, and the credit and honour of the
work done at Rothamsted on the nitrifying organisms belongs
entirely to Warington.

A few words must suffice for an account of the series of
Rothamsted experiments on animals.   Series II deals with
"Reports of experiments on the feeding of animals, sewage
utilisation, &c.   Published 1841—1895 inclusive," and contains
31 papers.   Among the points investigated may be mentioned—
the composition of foods in relation to respiration and the
feeding of animals; experiments on the feeding of sheep and
the fattening of oxen; some points in connection with animal
nutrition; the feeding of animals for the production of meat,
milk and manure.

The work on the part played by carbohydrates in the
formation of animal fat led to a keen controversy with foreign
investigators.   Lawes and Gilbert had satisfied themselves by

their experiments on pigs that fat was undoubtedly produced from carbohydrates. The German physiologists doubted this, and for some time there was a wordy warfare between the rival camps. Gradually the experimental evidence for the formation of fats from carbohydrates became overwhelming, and once again the Rothamsted position was vindicated.

Gilbert maintained throughout his life a close connection with foreign workers, and his holidays were frequently employed in visiting institutions and attending scientific meetings on the Continent. He made three visits to the United States and Canada and delivered several lectures there.

As he passed into old age his powers seemed to suffer little diminution, and his appearance at the age of eighty showed little indication of physical weakness. The death of Sir John Lawes in August 1900 was a severe blow to him, and soon afterwards his energies began to fail. He had a severe illness whilst away in Scotland in the autumn of 1901, but he recovered sufficiently to be able to return to his work for a short time. With the indomitable tenacity which had characterised him throughout life he continued actively at work for a few more weeks, eventually succumbing on December 23rd, 1901, in his eighty-fifth year.

Thanks are due to Dr J. A. Voelcker for kind assistance; and to the Royal Agricultural College Students' Club, Cirencester, for permission to reproduce the accompanying photograph.

Plate XX

HENRY WITHAM OF LARTINGTON

WILLIAM CRAWFORD WILLIAMSON
1816—1895

By DUKINFIELD H. SCOTT

Early exponents of Fossil Botany—Witham of Lartington—Edward William
Binney—William Crawford Williamson—early influences—first contribution to science—studies medicine—work on Foraminifera—appointed
Professor at Manchester—successful popular lecturer—his influence in
Natural History—investigation of the Carboniferous Flora—controversy
with French palaeo-botanists—the magnitude of his output—defects in
his work—later work at Kew—personal traits.

DURING the last forty years the study of fossil plants has
come to be a specially vigorous and characteristic branch of
British botany. The proper subject of my lecture is Williamson,
the man to whom above all others the present strong position
of the subject is due. But "there were brave men before
Agamemnon," and there are two of the older masters, Witham
and Binney, whom I cannot wholly pass over. I ought really
to include others, and notably Sir Joseph Hooker, to whom we
owe our first clear understanding of *Stigmaria* and of *Lepidostrobus*, but this course does not extend to those who, like Sir
Joseph, are still living among us and still in active work[1].

I am indebted to Mr Philip Witham, a member of the family,
for some information about Henry Witham, of Lartington, the
first Englishman to investigate the internal structure of fossil
plants.

Henry Witham was, by birth, not a Witham, but a Silvertop,
having been the second son of John Silvertop of Minster Acres,

---

[1] Since these words were spoken the veteran leader of English Botany has passed
away. A notice of Sir Joseph's career will be found in this volume, and the present
writer has given some account of his work on fossil plants in an *Anniversary Address
to the Linnean Society*, May 24th, 1912.

Northumberland. As Henry Silvertop he came in for the
Lartington property. He was born in 1779 and married Miss
Eliza Witham, niece and co-heiress of William Witham of Cliffe,
Yorkshire, when he took the name and arms of Witham.

The method of cutting thin sections of rocks and fossils
had just been invented by Nicol, and this gave Witham the
opportunity for his investigations. His papers are illustrated
by the botanist McGillivray, to whom he may have owed some
further assistance. Indeed he made little pretension to botanical
knowledge, but the opinions which he expresses strike one as
remarkably sensible, and he must have been a man of sound
judgment, at least in scientific affairs.

Witham was the first investigator of that most famous of
fossils, *Lepidodendron Harcourtii*; of the Craigleith tree (now
Pitys Withami), of the Lennel Braes trees (*Pitys antiqua* and
P. primaeva), of the Wideopen tree (*Pinites*, now *Cordaites
Brandlingi*) and of *Anabathra pulcherrima*. It is curious to
notice that the Craigleith tree, a manifest Gymnosperm, was at
first (1829) regarded even by the great Brongniart as a Mono-
cotyledon, while others imagined it to be a Lycopod. Witham,
however, soon set this right. He always speaks with great
respect of Brongniart, then just becoming the recognised leader
of fossil botany. The following passage from Witham's memoir
on the vegetable fossils found at Lennel Braes, near Coldstream,
is of interest.

"Now, according to that gentleman's [Brongniart's] opinion,
out of six classes...only two existed at that period [Carboniferous],
namely the Vascular Cryptogamic plants, comprehending the
Filices, Equisetaceae and Lycopodeae, and the Monocotyledons,
containing a small number of plants which appear to resemble
the Palms and arborescent Liliaceae. The existence, therefore,
of so extensive a deposit of Dicotyledonous plants, at this early
period of the earth's vegetation, appears to demand the attention
of the naturalist."

Brongniart's "Monocotyledons" were no doubt Cordaiteae.
Witham, we see, set the great man right as regards the antiquity
of Dicotyledons, in which, of course, Gymnosperms were then
included.

Witham's earlier papers were embodied in his book: *The Internal Structure of Fossil Vegetables found in the Carboniferous and Oolitic deposits of Great Britain, described and illustrated*, 1833. It is dedicated to William Hutton, author, with Lindley, of the *Fossil Flora* of Great Britain.

A passage from the dedication shows that Witham took his work seriously—"To lend my aid in bringing from their obscure repositories the ancient records of a former state of things, with the view of disclosing the early and mysterious operations of the Great Author of all created things, will ever be to me a source of unalloyed pleasure."

Witham thus fully realised the important significance of the work on which he was engaged. He must have been an interesting person of a somewhat complex character, and I wish we could know more about him. He died on Nov. 28th, 1844. Like all his family, he was a Roman Catholic[1].

Witham's localities on the Tweed remained practically unvisited until Mr Kidston re-explored them eight or nine years ago, with brilliant success—the results, however, are still unpublished.

Edward William Binney, the first investigator of the Lancashire coal-balls, was born at Morton in Nottinghamshire in 1812, and was thus only four years senior to Williamson. He settled in Manchester in 1836, and practised as a solicitor. He early showed scientific tastes; the Manchester Geological Society was started, chiefly by his influence, in October 1838. He was concerned in the discovery of the famous St Helen's trees, which first proved the connection between *Sigillaria* and *Stigmaria*. "Binney completed the proof that all coal-seams rest on old soils which are constituted entirely of vegetable matter; this was the seat-stone of a seam of coal" (Robert Hunt). He gave up the practice of Law, and, devoting himself to science, became a leading authority on northern geology, and rendered important aid to the Geological Survey by his

---

[1] The portrait of Henry Witham is from the original picture in the possession of the Salvin family, at Croxdale; a photograph of the picture was kindly obtained for me by Mr Philip Witham.

long experience of the coal-fields of Lancashire and Cheshire. He assisted in the discovery of the Torbane Hill mineral or Boghead Cannel, a deposit once notorious as a subject of litigation, and more recently as a bone of scientific contention. Binney died on December 19, 1881. Etheridge said of him: "He was a man of the highest honour and remarkably outspoken; his sturdiness and strength of character being rarely equalled."

Binney was the discoverer of some now famous fossils, notably *Dadoxylon* (now *Lyginodendron*) *oldhamium*, and *Stauropteris oldhamia.* His best known work is the monograph, *Observations on the Structure of Fossil Plants*, in four parts, published for the Palaeontographical Society, from 1868 to 1875. Thus his work on coal-plants overlapped that of Williamson.

The first part is on *Calamites* and *Calamodendron*—the names are used in the old sense, for Binney kept up Brongniart's distinction, though apparently not convinced of its validity. In this memoir he described the "cone of *Calamodendron commune*," now known as *Calamostachys Binneyana.*

Part II, on *Lepidostrobus* and some allied cones, is remarkable for the demonstration of heterospory in several species.

Part III, on *Lepidodendron*, deals partly with stems referred to *L. Harcourtii*, but now separated as *L. fuliginosum.* He also describes the structure of a *Halonia* and is led to the conclusion that it is the root of *Lepidodendron*. This view has not found favour, but our old ideas about *Ulodendron* and *Halonia* have been so upset of late, that everything seems possible!

Part IV is on *Sigillaria* and *Stigmaria*, the "*Sigillaria*" described being *S. vascularis*, since identified with *Lepidodendron selaginoides*, or *L. vasculare*, if we maintain Binney's specific name.

Binney was not a great theoriser. His object was rather to provide material for the botanists, he being essentially a geologist. This he did admirably, for his monograph is illustrated by magnificent drawings from the hand of Fitch, the famous botanical artist.

Binney stood more under the influence of Brongniart than did his successor Williamson.

Plate XXI

WILLIAM CRAWFORD WILLIAMSON (1876)

I now go on to my principal subject. Williamson's father, John Williamson, originally a gardener, was well known for his researches on the Natural History of the Yorkshire coast, and was for 27 years curator of the Scarborough Museum. Previously to that, John Williamson kept a private museum of his own, and it was in the room next to this that William Crawford Williamson was born on November 24, 1816. John Williamson's cousin, William Bean, was also an active local naturalist, known especially for his work on the Yorkshire Fossil Flora; the genus *Beania* is named after him.

Our Williamson's mother, born Elizabeth Crawford, was the eldest of 13 children of a Scarborough jeweller and lapidary. Young Williamson used to spend much time in the Crawford's workshop, watching them cutting and working with the diamond the agates from the gravels of the coast. "A youthful training," he says, "which became of the utmost value to me more than a third of a century later, when scientific research required me to devote much of my own time to similar work[1]."

In 1826 the famous William Smith and his wife established themselves in the Williamson's house, and stayed there for two years. Williamson's early recollections of the "Father of English Geology" must have been inspiring. His father was also a friend and correspondent of Sir Roderick Murchison.

The appearance of Phillips' classic volume, *Illustrations of the Geology of Yorkshire*, in 1829, gave young Williamson his first introduction to true scientific work. His father at once set to work to name from this book the fossils he collected, and his son was called in to help. "My evenings throughout a long winter were devoted to the detested labour of naming these miserable stones."—"Pursuing this uncongenial task gave me in my 13th year a thorough practical familiarity with the palaeontological treasures of Eastern Yorkshire. This early acquisition happily moulded the entire course of my future life[2]."

Those were not the days of the half-educated. Young Williamson, in addition to his special scientific training, had the advantage of a classical education, at schools both in

[1] *Reminiscences of a Yorkshire Naturalist*, p. 6.
[2] *Reminiscences*, p. 12.

England and France. The French part of his education was
not altogether a success, for most of the boys at the school were
English.

Passing through London on his return he had breakfast with
Sir Roderick Murchison, who took him to the Geological Society.
This was in March 1832, when he was little more than 15.
Certainly his entrance into the scientific world was made easy
for him. Would it be made equally easy now for a boy in
a similar position? In the same year, 1832, Williamson was
articled to Mr Thomas Weddell, a medical practitioner at
Scarborough. While with him, he continued to pursue Natural
History as a recreation—bird-collecting for example, and also
botany. He writes, "I was then forming a collection of the
plants of Eastern Yorkshire, as well as trying to master the
natural classification, which was already beginning to supplant
the Linnean method, so long the one universally adopted[1]."

A memoir on the rare birds of Yorkshire was communicated
to the Zoological Society of London—an early work though
not quite the earliest. While with Mr Weddell, Williamson
contributed a number of descriptions and drawings of oolitic
plants to Lindley and Hutton's *Fossil Flora*. He tells us how
the drawings had to be made in the evenings on Mr Weddell's
kitchen table. The plants he illustrated had for the most part
been collected by his father and John Bean in a small estuarian
deposit at Gristhorpe Bay. More than 30 species were thus
recorded by him.

He also made diagrams to illustrate some lectures on
Vegetable Physiology given by Mr Weddell at the Mechanics'
Institution. It is rather surprising to find that such a course
was given in a country town during the early 'thirties. Pro-
bably the learning displayed was not very deep, for Mrs Marcet's
Conversations seem to have been the chief authority.

In 1834–36 Williamson published important papers, deter-
mining geological zones, from the Lias to the Cornbrash, by
means of their fossils; subsequently he extended his zoning work
up to the Oxford Clay.

The opening of the Gristhorpe tumulus in July 1834, when

---

[1] *Reminiscences*, p. 33.

a skeleton, of the Bronze Age, was found in a coffin fashioned out of the trunk of an oak-tree, gave occasion to Williamson's one contribution to archaeology. His memoir was reprinted in the *Literary Gazette* for October 18, 1834 (still before he was 18). This was through Dr Buckland's influence; in a letter to Williamson he said, "I am happy to have been instrumental in bringing before the public a name to which I look forward as likely to figure in the annals of British Science." A second and third edition of this paper were called for.

In September 1835 Williamson was appointed curator of the Museum of the Manchester Natural History Society, and so began his long connection with the great northern town, lasting down to 1892. In those days the interest in the vigorous young science of geology was extraordinarily keen, and there was great activity, especially among the naturalists of the North, many of whom were working men. Williamson, about 1838, gave a course of lectures on geology at various northern towns, and thus raised funds for his removal to London, to continue his medical studies. It is interesting to find that Williamson, while at Manchester, helped to nurse John Dalton in his last illness.

While curator at Manchester, Williamson saw the rise of Binney as a geologist.

His remarks on the local study of botany at that time are interesting. "The botanical interests of the district were chiefly in the hands of the operative community. The hills between Lancashire and Yorkshire swarmed with botanical and flori-cultural societies, who met on Sundays, the only day when it was possible to do so[1]." Some of these men must have had an excellent education, as shown by the good English they wrote, as for example Richard Buxton, a poor working man, author of a standard *Botanical Guide*. The society to which Buxton belonged had, in 1849, existed for nearly a century. It may be doubted whether an equal enthusiasm for science still prevails in that or in any part of England.

In September 1840 Williamson went to London to complete his medical training, and entered University College, making

[1] *Reminiscences*, p. 78.

the acquaintance of Prof. Lindley, who had for so long known him only as a correspondent and collaborator.

Soon afterwards he was offered the post of naturalist to the Niger expedition, which he refused, and, as it turned out fortunately, for the journey proved disastrous. Stanger, of *Stangeria* fame, took his place.

In 1842, having then returned to Manchester and started in practice, Williamson made his first attempt at microscopic work, having become interested in the Foraminifera of the Chalk. He also began to examine Confervae, Diatoms and Desmids, finding perhaps, as others have done, that the Freshwater Algae give the best introduction to microscopic biology.

The work on Foraminifera became one of the most important in Williamson's career. In 1845 he wrote his valuable paper on microscopic organisms in the mud of the Levant. His work in this field culminated in his monograph of Foraminifera, issued by the Ray Society in 1857.

In 1851 Williamson was appointed Professor of Natural History, which included Zoology, Botany and Geology, at the new Owens College, Manchester. He tells us, "The botanical portion of my work was that for which I was least prepared" —"of the German language I was utterly ignorant[1]." The almost insuperable difficulties of a triple Professorship were at first met by spreading the complete course over two years, a sensible plan which was rendered impracticable by the more rigid requirements of examinations. It was not, however, till 1872 that a division of the duties of the chair took place; Williamson was then relieved of the geological teaching by the appointment of Prof. Boyd Dawkins; in 1880 the zoology was taken over by the late Prof. Milnes Marshall, Williamson thus retaining the very subject, botany, with which he had originally been the least familiar.

In addition to his peculiarly arduous duties as Professor, Williamson was a great populariser of science. He was one of the first two members of the Owens' staff to start, in 1854, evening classes for working men. He gave numerous scientific lectures at the Royal Institution in London and elsewhere, his

[1] *Reminiscences*, p. 136.

Plate XXII

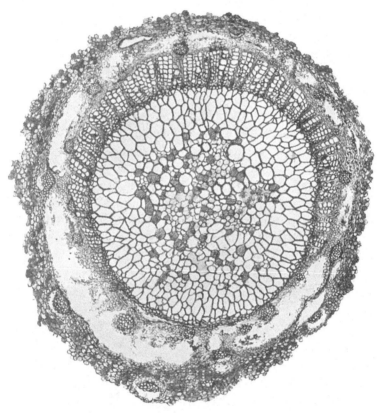

Vascular system of stem of *Lepidodendron selaginoides*
in transverse section

Drawn by Williamson

greatest work in this field being his lectures for the Gilchrist Trustees. He mentions that from 1874 to 1880 he delivered 158 of these lectures in 61 towns, and he continued this work with equal activity for another 10 years. He was a vigorous and effective lecturer, who always interested his audience; he illustrated his lectures by bold diagrams, drawn by his own hand. In order to form any idea of Williamson's many-sided activity it must be remembered that he was all the time engaged in active medical practice, both general and special, for he was well known as an aurist. Yet he always found time for fruitful original research, often of the most laborious character.

Prof. Judd says, in a letter written to me in February 1911:

"I have often been struck by the fact that Williamson, appointed to an impossible Professorship of Zoology, Botany and Geology, managed to initiate *great movements* in connection with *each of these sciences.*

"In Geology he was clearly the pioneer in the subdivision of formations into zones each characterised by an assemblage of fossils—Ammonites playing the most important part....But Williamson did another great service to Geology....Sorby visited Williamson at Manchester and learned the art of making sections which he applied with such success to the study of igneous and other rocks, becoming the 'Father of Micropetrography.'

"In Zoology, Williamson initiated the work done in the study of *deep-sea deposits*, by his remarkable memoir on the mud of the Levant, in 1845, when he was 29 years old. This led to his study of the Foraminifera (especially by the aid of thin sections) and to his monograph in the Ray Society on that group....

"Of his contributions to Botany through his sections of 'Coal balls' I need say nothing."

Prof. Judd makes no reference here to the papers which obtained for Williamson his F.R.S. in 1854. These embodied his researches on the development of bone and teeth, in which he demonstrated that the teeth are dermal appendages homologous with the scales of fishes. This important work dated back to 1842 and was inspired by his enthusiasm for the then novel cell-theory of Schleiden and Schwann.

The interest aroused by this investigation is shown by the fact that the great German anatomist Kölliker travelled to Manchester, about the year 1851, to see Williamson's preparations.

As regards Williamson's work as a botanist, in which we are chiefly interested in this course, his best contribution to recent botany was no doubt his investigation of *Volvox*, published in 1851 and 1852, in which he traced the development of the young spheres and the mode of connection of their cells, anticipating the results of much later researches.

He was a great lover of living plants; his garden and greenhouses at Fallowfield, his Manchester home, were of remarkable interest, and he was a keen gardener. At the British Association Meeting of 1887 one of his guests said that "most of the distinguished botanists of Europe and America were in the garden, and not one but who had seen something growing he never saw before[1]." Insectivorous plants and the rarer vascular cryptogams were specially well represented. It was from his private garden that his classes were supplied with specimens.

As we have seen, fossil plants engaged Williamson's attention in his earliest years, when as a mere boy he contributed to Lindley and Hutton's *Fossil Flora*.

His first important independent work in this field was his paper "On the Structure and Affinities of the Plants hitherto known as Sternbergiae" (1851), in which he proved, for the first time, that these curious fossils, resembling a *rouleau* of coins, were casts of the discoid pith of *Dadoxylon*, or, as we should now say, of Cordaiteae—the first step in the reconstruction of this early gymnospermous family. This investigation, to which he appears to have been led almost accidentally, through some good specimens coming into his hands, brought him back, as he says, to his old subject of fossil botany. It was long, however, before he got fairly started on his great course of investigations on Carboniferous plants.

In the meantime he had returned to the Yorkshire Oolitic plants and, about 1847, published a paper in the Proceedings of the Yorkshire Philosophical Society, "On the Scaly Vegetable Heads or collars from Runswick Bay, supposed to belong to the

[1] *Reminiscences*, p. 190.

Zamia gigas." His full paper, in which he maintained the Cycadean affinities of the flower-like fossils, was written soon afterwards, but met with a series of misfortunes, and was not finally published till 1870, in the *Transactions of the Linnean Society*, before which body it had been read in 1868. Williamson was admittedly right in connecting the floral organs with the so-called *Zamia* foliage, and his interpretation of the complicated structure was as good as was possible in the then state of knowledge. The true nature of these fossils, now known by the name *Williamsonia*, given them by Mr Carruthers, could only be understood at a much later date in the light of Dr Wieland's famous researches on the American Bennettiteae, and has quite recently been made clear in a memoir by Prof. Nathorst. Perhaps, even now, some points remain doubtful.

Early in the fifties Williamson made some rough sections of a Calamite which came into his hands, and this was the beginning of his most characteristic line of work. A remarkable internal cast of a Calamite, figured by Lyell in his *Manual of Geology* in 1855, led to a correspondence with M. Grand'Eury, now so famous as the veteran French palaeobotanist. Williamson at that time had no intention of entering on the serious study of Carboniferous plants, for Binney was already in the field. Grand'Eury's letter, however, caused him to look up his old sections, which he found differed from the Calamitean stems figured by Binney. Matters for a time moved slowly, and Williamson's specimen was only described in 1868 in the *Manchester Memoirs*. This fossil, which he named *Calamopitus*, is now known as *Arthrodendron*, and is a distinct type of Calamarian stem, intermediate between the common *Calamites* or *Arthropitys*, and the more elaborate *Calamodendron* of the Upper Coal Measures.

Williamson was now fairly started on his Carboniferous work. His first memoir on the Organisation of the Fossil Plants of the Coal Measures was communicated to the Royal Society on November 11, 1870. It is amusing to find that the secretaries objected to the memoir being called Part I, since it bound the society to publish a Part II! Nineteen Parts were published, the last in 1893.

The first memoir was on the Calamites, and controversy at once broke out. Williamson was from the first impressed by the manifest occurrence of exogenous, or, as we should now call it, secondary growth, both in the Calamites and the Lepidodendreae, groups which he was convinced were cryptogamic. The controversy with the great French school, headed by the illustrious Brongniart, is well known. As Williamson put it: "The fight was always the same; was Brongniart right or wrong when he uttered his dogma, that if the stem of a fossil plant contained a secondary growth of wood, the product of a cambium layer, it could not possibly belong to the cryptogamic division of the vegetable kingdom?[1]"

In England, however, the dispute was on different lines. "In August of 1871," says Williamson, "the British Association met at Edinburgh. At that meeting I brought forward the subject of cambiums and secondary woods in Cryptogams, with the result that my views were rejected by every botanist in the room." There followed a controversy in the pages of *Nature*, which is of some interest, as showing the state of opinion in England at that time. Williamson tells us in his autobiography the principle by which he was guided in his work: "I determined not to look at the writings of any other observer until I had studied every specimen in my cabinet, and arrived at my own conclusions as to what they taught." In spite of this excellent rule it is probable that he was at first unconsciously influenced by the views of Brongniart, which may have led him to attach too much *systematic* importance to the occurrence of secondary growth. At any rate he proposed at the Edinburgh meeting "to separate the vascular Cryptogams into two groups, the one comprehending Equisetaceae, Lycopodiaceae and Isoëtaceae, to be termed the Cryptogamiae Exogenae, linking the Cryptogams with the true exogens through the Cycads; the other called the Cryptogamiae Endogenae, to comprehend the Ferns, which will unite the Cryptogams with the Endogens through the Palmaceae[2]."

It is curious to note in passing that his main divisions, so far as vascular Cryptogams are concerned, correspond to the

---

[1] *Reminiscences*, p. 203.   [2] *Nature*, Vol. IV., 1871, p. 357.

Plate XXIII

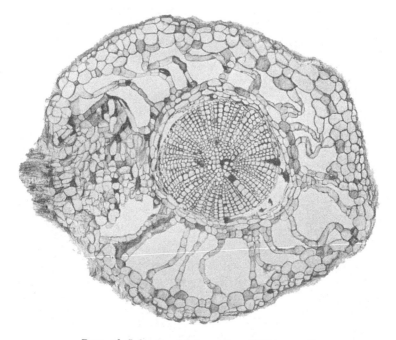

Root of *Calamites* (*Astromyelon Williamsonis*)
in transverse section

Drawn by Williamson

Lycopsida and Pteropsida of Prof. Jeffrey, though the suggested relation to the higher plants would not be accepted by any modern botanist. In spite of Williamson's tactical error in weighting himself with a doubtful scheme of classification, and in spite also of a faulty terminology, it is easy to see now that he had the best of the controversy, for he knew the facts about the structure of the Carboniferous Cryptogams, which his opponents, at that time, did not. They stuck to generalities, and those who take the trouble to rake the ashes of this dead controversy will at least learn that dogmatism is not confined to theology!

An interesting point is that Williamson at that time spoke of Brongniart almost as an ally[1]. The conviction that the old Lepidodendrons and Calamites were "exogenous" then seemed to him of greater importance even than his belief that they were Cryptogams. The English opposition, however, was never really formidable, and so a change of front became necessary, to meet the attacks of the powerful French school. Williamson was an energetic disputant; not content with his numerous English publications, he published, in 1882, an article in the *Annales des Sciences Naturelles*, entitled "Les Sigillaires et les Lepidodendrées." This was translated into French for him by his colleague Marcus Hartog, whose assistance he greatly valued. He describes this vigorous polemical treatise as "flung like a bombshell among my opponents."

In time they came over, one by one, to his views, and even the most redoubtable of the French champions Bernard, Renault, before the close of his life, had made very considerable concessions to Williamson's side of the question. There is no need to dwell on the controversy; every student now knows that the Club-mosses, the Horse-tails and the Sphenophylls of Palaeozoic times formed abundant secondary tissues homologous with those of a Gymnosperm or a Dicotyledon; the case of the Sphenophylls shows that the character was not limited to arborescent plants then any more than it is among Dicotyledons at the present day. At the same time, as Williamson maintained, these groups of plants were, broadly speaking, cryptogamic.

[1] *Loc. cit.*, p. 409.

On the other hand it has been said by a distinguished botanist that in the Fern-series secondary growth came in together with the seed. This is not strictly correct, but it is true that the plants such as *Lyginodendron*, which Williamson in his later publications cited as Ferns with secondary growth, have turned out to be seed-bearing. Even among the Lycopods a certain proportion of the Lepidodendreae bore organs closely analogous to seeds. These partial concessions, which may now gracefully be made to the old Brongniartian creed, do not however really affect the importance of Williamson's results, which Count Solms-Laubach has well summed up in the following words: "It was thus made evident by Williamson that cambial growth in thickness is a character which has appeared repeatedly in the most various families of the vegetable kingdom, and was by no means acquired for the first time by the Phanerogamic stock. This is a general botanical result of the greatest importance and the widest bearing. In this conclusion Palaeontology has, for the first time, spoken the decisive word in a purely botanical question[1]."

To attempt a review of Williamson's work in fossil botany would be to write a treatise on the Carboniferous Flora. In every group—Calamites, Sphenophylls, Lycopods, Ferns, Pteridosperms, Gymnosperms—his researches are among the most important documents of the palaeobotanist, and to a great extent constitute the basis of our present knowledge. At the time he wrote, the wealth of his material was absolutely unrivalled, and its abundance was only equalled by the astonishing energy and skill with which he worked it out.

As regards the Calamites, he demonstrated, to use his own words, "the unity of type existing among the British Calamites," abolishing the false distinction between Calamiteae and Calamodendreae.

Among the Sphenophyllums (although there was at first some confusion in his nomenclature) he gave the first correct account of the anatomy, and of the organization of the cone.

Concerning the Lycopods, the greater part of our knowledge is due to him. He described the structure in ten species referred

[1] *Nature*, Vol. LII. 1895, p. 441.

Plate XXIV

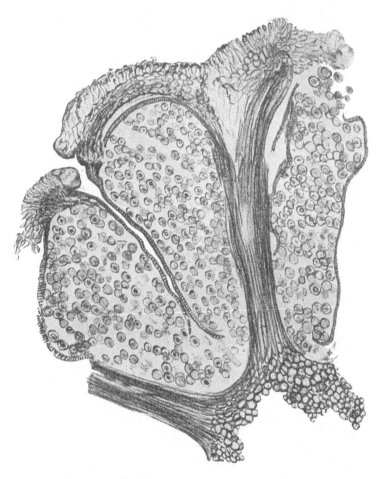

Cone of *Calamostachys Binneyana* ; sporangia and sporangiophores

Drawn by Williamson

to *Lepidodendron*, besides other allied forms, and placed our knowledge of the comparative anatomy, once for all, on a broad and secure basis. His great monograph of *Stigmaria*, by some considered his best work, is still our chief authority for the subterranean organs.

In the Ferns he made important contributions to our knowledge of the group now familiar to botanists as the Primofilices of Arber. In particular his account of the plant now known as *Ankyropteris corrugata* is still among the best we possess of any member of the family.

In Pteridosperms, to use the modern name, Williamson may fairly be called the discoverer of the important family Lyginodendreae. He appreciated their intermediate position, speaking of them, in 1887, as "possibly the generalised ancestors of both Ferns and Cycads."

As regards both Pteridosperms and Gymnosperms proper, attention may be specially called to his work on isolated seeds, in which he was surpassed by Brongniart alone. This field of investigation, long neglected, has lately been revived with striking results.

I hope that all students of fossil botany will have at least turned over the pages and the plates of Williamson's works, for only by inspection of the original memoirs can any idea be gained of his vast services to our science.

His remarkable skill as a draughtsman (for all his memoirs are illustrated by his own hand) is not always done justice to in the published reproductions as the fine examples of his original drawings, so kindly lent for the lecture by Mrs Williamson, will show[1]. At the time when Williamson's main work was in progress—from 1870 to 1892—geologists were probably more appreciative of its value than botanists. Happily, in spite of occasional trouble with Referees, none of his work was lost, the Royal Society going steadily through with all the nineteen memoirs which were entrusted to them.

The one botanist, who, up to the year 1890, estimated Williamson's work at its full value was Count Solms-Laubach,

---

[1] Three characteristic figures from these originals have been reproduced for this volume (Plates 22-24).

who makes the honourable boast that he knew Williamson's collection as no one else did.

Williamson's writings are not easy reading, especially for the modern botanical student, for the terminology is often unfamiliar, and the arrangement of the matter unsystematic. It would be out of place to enter on a criticism of details, but it is necessary to call attention to the one serious mistake which ran through much of Williamson's work, though at the last he to a great extent corrected it himself. He was always too ready to interpret specimens of the same fossil plant which differed in size and anatomical complexity, as developmental stages of one and the same organ. Such differences among fossils are more often due to the order of the branch on the plant, or to the level at which a section is cut. This error led to some mistaken, and indeed impossible views of the process of development. I mention this partly because I have noticed the same fundamental mistake in the work of much more modern writers. "We are none of us infallible—not even the youngest of us," and among the latest fossil-botany papers I have read, I have detected the very same confusion between differences of size and differences of age, which constitutes the most serious blemish in Williamson's writings.

As is well known, Williamson in his latest independent work corrected, as regards the Lepidodendrons, on the basis of a laborious re-investigation, the chief mistake he had made as to their process of growth[1]; he thus displayed an openness of mind worthy of a great naturalist.

I first saw Williamson on February 16, 1883, when I attended his Friday evening lecture at the Royal Institution, "On some anomalous Oolitic and Palaeozoic Forms of Vegetation." I did not, however, make his acquaintance till six years later, when we met at the British Association Meeting at Newcastle-upon-Tyne, in 1889. This led to a visit to his house in company with Prof. Bower; it was on March 8, 1890, that

---

[1] Williamson, "On the light thrown upon the question of the Growth and Development of the Carboniferous Arborescent Lepidodendra by a study of the details of their Organisation." *Mem. and Proc. Manchester Lit. and Phil. Soc.*, Ser. IV. Vol. IX. 1895.

I first had a sight of his collection. I find the entry in my diary: "Spent 7 hours over fossils, especially *Lyginodendron* and *Lepidodendron*, preparations magnificent." I at once became an ardent convert to the cult of fossil plants to which I had hitherto been indifferent, though I must in fairness admit that Count Solms-Laubach's *Einleitung* had done something to prepare the way. I well remember the state of enthusiasm in which I returned home from Manchester. A subsequent visit confirmed me in the faith, but it was some little time before I put my convictions into practice. In 1892 Williamson, then in his 76th year, resigned the Manchester Professorship and came to live near London. In the same year I migrated to Kew, and it was agreed that we should work in concert, an arrangement which received every encouragement from the then Director, Thiselton-Dyer. Williamson first came to the Jodrell Laboratory on Friday, December 2, 1892. Then, and on many later visits, he carried a satchel over his shoulder, crammed with the treasures of his collection. For some months he came pretty regularly once a week, afterwards less often. On these visits we discussed the work I had done on the sections during the interval, and sometimes our discussions were decidedly lively. In the end, however, we always managed to come to a satisfactory agreement. Our first joint paper (*Calamites, Calamostachys* and *Sphenophyllum*) was sent off to the Royal Society, rather more than a year from the start, on December 29, 1893.

During the early part of 1894 Williamson came occasionally to Kew, and our discussions were renewed, this time chiefly on *Lyginodendron*. Our second paper (Roots of Calamites) was despatched on October 30, 1894.

After a considerable interval Williamson again visited Kew, on December 12, 1894, when we started on his *Lepidodendron* sections, a subject on which we never published in conjunction. His last visit was on January 7, 1895. A few days later his health broke down, and though there were many fluctuations he was never able to come to the laboratory again. I saw him last, at his own house, on June 4th. On the 13th I read our joint paper on *Lyginodendron* and *Heterangium* at the Royal Society; on the 23rd he passed peacefully away.

If Williamson could have lived it would, I think, have given him great pleasure to see the success, in his own country, of the work which he inaugurated and the progress of the subject to which he devoted the last 25 years of his life. I am happy to believe that he felt in the evening of his days, that the period of comparative neglect through which his work had passed, was at an end. For myself, I may say that my work, since I knew Williamson, owes its inspiration to him. But quite apart from our scientific relations it is a great privilege to have known him. Though his many-sided activity, as physician, professor, popular lecturer, geologist, zoologist, botanist and artist involved an amount of work which to us of a less strenuous generation is almost inconceivable, Williamson was as far as possible from being the mere student. His personality was intensely human. He was a man of most decided likes and dislikes; his conversation was often brilliant, and sometimes vigorous to an almost startling degree.

The grand old race of all-round naturalists found in Williamson its worthy culmination, and we can only regret that, from the nature of the case, he can have no equal successor[1].

[1] The portrait of Williamson is from a photograph kindly lent by Mrs Williamson, and taken, as she informs me, at Torquay in or about 1876, when he was about 60.

Plate XXV

HARRY MARSHALL WARD (1895)

HARRY MARSHALL WARD
1854—1906

By Sir WILLIAM THISELTON-DYER

Training at South Kensington—Cambridge—Germany—investigates coffee disease in Ceylon—his early investigations—appointment to Manchester and association with Williamson—Ward's brilliance as an investigator —Cooper's Hill—investigation of lily disease—leguminous root tubercles — symbiosis and the ginger-beer plant — the Croonian Lecture —the bacteriology of water—bactericidal action of light—Ward's "law of doubling"—appointment to Cambridge—mycopiasm controversy—infection and immunity—physiological varieties of Rusts—bridgeing species—illness and death—his record as an investigator—personal characteristics.

HARRY MARSHALL WARD, eldest son of Francis Marshall Ward, was born in Hereford, March 21, 1854, but he came of a Lincolnshire stock, settled for some time in Nottingham. From unavoidable causes he left school at 14, but afterwards continued his education by attending evening classes organised under the Science and Art Department. To that Department, he owed indirectly the opportunity of a useful and brilliant career. His means were small, and his earliest aim was to qualify as a science teacher. He was admitted to a course of instruction for teachers in training given by Prof. Huxley in 1874–5. Although he must have derived from it a sound insight into the principles of zoology, the subject does not seem to have had any permanent attraction for him.

In the summer of 1857 Ward came under my hands in a course of instruction in botany which I conducted with Prof. Vines in the Science Schools at South Kensington, and from this time onwards we were in intimate relations to the close of his life. I can best tell the story as it came under my eyes.

It contains much that could not easily be dealt with in any other way.

It was soon apparent that we had got hold of a man of exceptional ability. It must be confessed that the atmosphere was stimulating, and the conditions under which the teaching was carried on necessitated its being given at high pressure. I remember that on one occasion Ward fainted at his work, from no other cause, I think, than over-excitement. In the autumn of the same year he went for one session to Owens College, Manchester, with the object of continuing his general education. I learn that he carried off the prizes in every subject that he took up.

In the succeeding year I was glad to avail myself of the assistance of Ward as demonstrator in a subsequent course at South Kensington, which I undertook with Prof. Vines. Later in the year he became a candidate for and secured an open scholarship at Christ's College, where Vines himself was then a Fellow, and went into residence in October, 1876.

Ward took full advantage of his opportunities at Cambridge, and attended the teaching of Sir Michael Foster in physiology and of Prof. F. M. Balfour in comparative anatomy. The sound and fundamental conceptions which he acquired from the former manifestly influenced his work throughout life. He took a first class in botany in the Natural Science Tripos in 1879. His first published paper was the result of work in the same year in the Jodrell Laboratory at Kew. In this, which was published in the *Proceedings of the Linnean Society*, he seriously criticised and corrected that of Vesque on the embryo-sac of Phanerogams.

As was customary with our young botanists, Ward went to Germany for a short time, for purposes of study and to strengthen his knowledge of the language. He worked at Würzburg with Sachs, whose lectures on the physiology of plants he afterwards translated in 1887. There he continued his study of the embryo-sac in Orchideae, as Sachs subsequently testified, "zu meiner vollsten Zufriedenheit."

Before the end of the year Ward was appointed on the recommendation of Kew to proceed to Ceylon for two years as Government Cryptogamist to investigate the leaf-disease in coffee. The history of this malady is almost unique in vegetable pathology. A native fungus which had eluded scientific observation, and must therefore have maintained an inconspicuous and limited existence on some native host-plant, found a wider opportunity on the Arabian coffee plant and fell upon it as a devastating scourge. It was first detected in 1869 on a single estate; in 1873 there was probably none in the island entirely free from it. Mr (since Sir Daniel) Morris had shown that the plants could be cleansed by dusting them with a mixture of sulphur and lime. But the remedy proved of no avail as the plants speedily became re-infected. Morris had been transferred to another appointment in the West Indies and Ward's duty was to take up the investigation. This he accomplished exhaustively. He showed that the fungus (*Hemileia vastatrix*) was one of the Uredineae and that infection was produced by the wind-borne uredospores. Had the planters, as in Southern India, left forest belts between their plantations, the spores might have been filtered out and the disease controlled. As it was it spread like an unchecked conflagration. Ward also discovered the teleutospores; nothing has been added to our knowledge of its life-history beyond what he obtained. The result of his investigations was given in three official reports and in papers contributed in 1882 to the Linnean Society and the *Quarterly Journal of Microscopical Science*. It was no blame to him that his work led to no practical result. The mischief admitted of no remedy. The coffee-planting industry of Ceylon was destroyed and the Oriental Bank succumbed in the general ruin. Leaf disease has now extended to every coffee-growing country in the Old World from Natal to Fiji.

In a tropical country leaves supply a substratum to a little flora of their own, consisting of organisms partly algal, partly fungal, in their affinity. Ward, who had already developed his characteristic habit of never neglecting any point incidental to a research, carefully studied them, in order both to ascertain how far their presence affected the health of the leaf itself and to work out their life-history. The outcome was three important papers. One on Meliola, an obscure genus of tropical epiphyllous fungi, belonging to the Pyrenomycetes, was published

in the *Philosophical Transactions* in 1883. Bornet's classical
memoir published in 1851 had been the authority on the subject.
Ward was able to fill up "large gaps in the knowledge of important details." Another paper published in the *Quarterly
Journal of Microscopical Science* in 1882 on an Asterina illuminates
an allied organism. But the crown of all Ward's Ceylon work
was the splendid memoir on a Tropical Epiphyllous Lichen
which was published by the Linnean Society in 1883. In this
he, I think, cleared up much that was obscure in the *Mycoidea
parasitica* described by D. D. Cunningham. Having myself
communicated the paper, I shall always remember the pleasure
with which I undertook in Ward's absence to give an account of
it. He solved the problem with convincing completeness; he
extended Schwendener's lichen theory to a group of obscure
epiphyllous organisms of which he afforded, for the first time, a
rational explanation. The success with which this was accomplished placed him at once in the first rank of mycological
investigators.

De Bary was the leading authority on Uredineae; and in
1882 Ward paid a short visit to him at Strasburg to confer with
him on his coffee disease work, the accuracy of which de Bary
entirely confirmed. There he made the acquaintance of Elfving
and completed his Meliola paper.

The outlook for Ward was now precarious. Fortunately,
I found myself sitting next to Sir Henry Roscoe at a Royal
Society dinner, and I suggested that Ward, as an old student of
Owens College, would be a fitting recipient of a Bishop Berkeley
Fellowship for original research. Principal Greenwood recorded
the fact that "the very important results already achieved by
Mr Ward in Ceylon, in the domain of the higher botany, led
the Senate and the Council to make this appointment." In
1883, he was appointed Assistant Lecturer and Demonstrator in
Botany, and, on the same testimony, "abundantly justified his
election." It was a peculiar pleasure to him to relieve the
veteran Professor Williamson by taking entire charge of
Vegetable Physiology and Histology. His position was, in the
same year, made secure by his election to a Fellowship at
Christ's College, and he married the eldest daughter of the late

Francis Kingdon, of Exeter, who was a connection of Clifford the mathematician.

The passion for research now completely possessed Ward and never left him for the rest of his life. He published papers which added much to our knowledge of the Saprolegnieae a group of fungi of aquatic habit, partly saprophytic and partly parasitic. It is interesting to note that he was particularly attracted by the mode in which the hyphae attack the tissues on which they prey. This was a matter on which he subsequently threw an entirely new light. He made the interesting discovery of an aquatic Myxomycete, such a mode of existence being hitherto unknown in the group, and worked out its life-history. But his mind had now become definitely fixed on the problems presented by plant diseases, and they remained the principal occupation of his life. In their widest sense these resolve themselves into a consideration of the mode in which one organism obtains its nutriment at the expense of another. This ranges from a complete destruction of the host by the parasite to a harmless and even advantageous symbiosis. He was thus naturally led to an exhaustive study of the literature of the Schizomycetes, and contributed an article on the group in 1886 to the *Encyclopaedia Britannica*, which, for the time at any rate, gives the best account of it, certainly in English, and probably in any other language. When he supplemented this in 1902 by the article on Bacteriology, it was largely to give an account of his own important discoveries. In the earlier one, he had pointed out the difficulties of a natural classification of Schizomycetes due to their pleomorphism, which Lankester had demonstrated in 1873. He returned to the subject in an article in the *Quarterly Journal of Microscopical Science* in 1892. It may be noted that, in his British Association address at Toronto, he took occasion to put in their proper relation the work of Cohn and of his pupil Koch.

In 1885, the Regius Professorship of Botany at Glasgow was vacant by the transference of Prof. Balfour to Oxford. Ward was a candidate with the warm support of his fellow-botanists. It was thought that his Colonial services would weigh with the Government; but other influences were at work in favour of another candidate, whom, however, the University refused to accept. A deadlock ensued, which was only solved by the Government finally refusing to appoint either candidate. This was a great disappointment to Ward, which was in some degree mitigated by his appointment to the new Chair of Botany in the Forestry Branch of the Royal Indian Engineering College, Cooper's Hill. The utilitarian atmosphere in which he found himself was not very congenial to him. But he had at any rate at last some sort of adequate position and a laboratory to work in, and here he remained—not, I think, unhappily—for ten years. He was, as he had been at Manchester, a successful teacher, and had the gift of interesting his pupils, whom he used to bring weekly to Kew during the summer months to visit the Arboretum. In point of research, this was the period of much of his most brilliant work.

The study of Uredineae occupied Ward at intervals during his life. The reproductive organs are pleomorphic, and it is no easy task to ascertain with certainty those that belong to the same life-history. In a paper on *Entyloma Ranunculi*, published in the *Phil. Trans.* in 1887, Ward for the first time traced the germination of the conidia of an Entyloma, and confirmed Winter's suggestion that they were not an independent organism, but actually belonged to it. Incidentally he discussed the conditions which are favourable to the invasion of a host by a parasitic fungus. This raised the question of immunity, to which at intervals he repeatedly returned.

About the same time he published in the *Quarterly Journal of Microscopical Science* the results of an investigation undertaken for the Science and Art Department on the mode of infection of the potato plant by *Phytophthora infestans*, which produces the potato disease. It was not easy to add anything to the classical work of de Bary, but it was ascertained that "the development of the zoospores is delayed or even arrested by direct daylight," and Ward's attention was attracted to the problem, which he afterwards solved, of how the hyphae erode the cell-wall.

The solution was given in 1888 in a paper in the *Annals of Botany*, "On a Lily Disease," which has now become classical.

He discusses the fungus which produces it, and shows that the tips of the hyphae secrete a cellulose-dissolving ferment which enables them to pierce the cell-walls of the host. This ferment has since been described as cytase. He shows that its production would determine the passage from a merely saprophytic to a parasitic habit, and makes the suggestion that an organism might be educated to pass from one to the other.

An admirable research (1887) was on the formation of the yellow dye obtained from "Persian berries" (*Rhamnus infectorius*). A dyer had found that uninjured berries afforded a poorer colouring liquor than crushed. Gellatly had found, in 1851, that they contained a glucoside, xanthorhamnin, which sulphuric acid broke up into rhamnetin and grape-sugar. The problem was to localise the ferment which did the work. Ward obtained the unexpected result that it was confined to the raphe of the seed.

As early as 1883 Ward had attacked a problem which he pursued at intervals for some years, and which was fraught with consequences wholly unforeseen at the time. It had long been known that leguminous plants almost invariably carried tubercular swellings on their roots. The opinion had gradually gained ground that they were due to the action of a parasite. Bacteria-like corpuscles had been found in the cells of the tubercle, and it was assumed that they had played some part in exciting the growth of the latter. "No one had as yet succeeded in infecting the roots and in producing the tubercles artificially." Ward described, in a paper in the *Phil. Trans.* in 1887, how he had accomplished this. He showed, in fact, that a definite organism invades the roots from the soil, and finds its access by the root-hairs.

Lawes and Gilbert had long ago proved that the higher plants are incapable of assimilating free nitrogen. Hellriegel and Wilfarth had, however, shown in 1886 that leguminous plants carry away more nitrogen from the soil than could be accounted for. This Ward confirmed by his own pot-experiments, and satisfied himself that the excess could only be derived from the free nitrogen of the air. Hellriegel further concluded that the tubercles played an essential part in the process. Ward had no doubt that the bacteroids were the channel of supply. But he failed to get any proof that they could assimilate free nitrogen outside the plant. He suggested that their symbiosis might be an essential condition, and was obliged finally to leave it an open question whether the cells of the tubercles or the bacteroids were the active agents in nitrogen assimilation. He had already stated in 1887 that it is very probable that the bacteroids " may be of extreme importance in agriculture." But he was never satisfied with anything short of the strictest proof.

In 1890 Ward was invited to deliver the Croonian Lecture. He chose for his subject the relation between host and parasite in plant disease. He defined disease in its most generalised form as " the outcome of a want of balance in the struggle for existence." But the particular problem to which he addressed himself was the way in which the balance is turned when one organism is invaded by another. This is the most common type of disease in plants and a not infrequent one in animals. The first result reached was identical with that of Pasteur for the latter ; the normal organism is intrinsically resistant to disease. It is an immediate inference that natural selection would make it so. Ward then discusses very clearly the physiological conditions of susceptibility, which he shows to be a deviation from the normal. He had already indicated this in the case of Entyloma. The epidemic phase is reached when the environment is unfavourable to the host but not so or even favourable to the parasite. He then attacks the more obscure case where there is no obvious susceptibility. This, he finds, resolves itself into a mere case of the struggle for existence : "a struggle between the hypha of the fungus and the cells of the host." It is more subtle in its operation but of the same order of ruthlessness as the ravages of a carnivore. Ward's account of the struggle is almost dramatic. The cellulose " outworks " are first broken down, as he had previously shown, by a secreted ferment. The " real tug of war " comes when the hypha is face to face with the ectoplasm. Its resistance is at once overcome by flooding it with a poison, probably oxalic acid.

War with attack and defence is a product of evolution. How did it come about in this particular case? Ward convincingly traces out the whole process. The normal plant obtains its food from inorganic material. But when opportunity offers it easily lapses into a condition in which it takes the material for metabolism ready made from the decay of others and becomes saprophytic. Ward shows that it is only a step to the attack on the living, and for the saprophyte to become a parasite, and he further shows that it can be readily educated to be so. He does not hesitate to suggest that the function of conidia in the complicated cycle of fungal reproduction is to form the cellulose-dissolving ferment. But now and again the host does not succumb to its invader. A truce is sometimes called in the struggle, and host and parasite are content to live together in a mutually advantageous symbiosis or commensalism.

Three years earlier, in 1887, Ward's attention had been drawn by a happy accident to the physiological aspect of symbiosis, and it never ceased to occupy his mind. It was well known that ginger-beer was made in villages in stone bottles. The fermentation was effected by the so-called "ginger-beer plant" which was passed on from family to family, but nothing was known as to how or where it originated. It seemed to have some analogy with the Kephir of the Caucasus. A specimen was sent to me from the Eastern Counties, and it stood for some time in the sun in my study. I noticed the vigorous growth accompanied by a copious evolution of gas. Ward coming to see me one day, I handed it over to him as a problem worth his attention. At the same time Prof. Bayley Balfour had examined it and concluded that it was a mixture of a yeast and a bacterium. Its study involved Ward in a very laborious research which occupied him for some years, and of which the results were published in the *Phil. Trans.* in 1892. It proved to be a mixture of very various organisms, every one of which Ward exhaustively studied. This required not less than 2000 separate cultures. The essential components proved to be, as Balfour had suggested, a yeast derived from the sugar and a bacterium from the ginger. Both were anaërobic; the yeast fermented cane-sugar with the copious production of carbon dioxide but little alcohol; the bacterium also produced carbon dioxide, even in a vacuum tube.

The action of the two components studied separately proved to be not the same as when they worked in concert. This was conspicuously the case with the evolution of carbon dioxide, which proceeded with such violence as to make the research attended with considerable danger. It is known that the action of ferments may be checked by the inhibition of the products formed. Ward pointed out that while the use of these might be advantageous to the bacterium, their consequent removal might be equally so to the yeast. This established the important principle of symbiotic fermentation and gave it a rational explanation. On the morphological side Ward showed that the ginger-beer plant is comparable to a gelatinous lichen, and, having resolved it into its constituents, successfully reconstituted it.

The new conception threw a flood of light on many obscure points in fermentation generally, and it is not surprising that Ward's work at once attracted the attention of the brewing industry. It led him to an even more fertile suggestion, that of metabiosis. It was known that the finest wine is sometimes produced from mouldy grapes. He regarded this as a case of one organism preparing the way for another. He returned to the subject in a lecture given at the British Association at Dover in 1899 and pointed out that in the Japanese manufacture of Saké, an Aspergillus prepares the way for the yeast. He also showed that metabiosis played an important part in nitrification.

Fungi cannot draw their nutriment from solid materials without first profoundly modifying them. They accomplish a large part of their digestion, so to speak, externally to themselves. This constantly occupied Ward's mind. He insisted on the part played in the process by ferments. The hyphae of Stereum (*Phil. Trans.* 1898) delignify the walls of the wood elements of Aesculus layer by layer, and then consume the swollen cellulose. He failed, however, to isolate the ferment which does the work. Nor was he more fortunate with the little known fungus Onygena, which grows on horn, hoofs and hair, setting free ammonia as a final product (*Phil. Trans.* 1899). That there must be some hydrolysis of keratin can hardly be doubted, for Ward established the remarkable fact that the walls of the hyphae contain no cellulose, but are composed of chitin. Onygena has, in fact, abandoned a plant for an animal nutrition. This would place the germination of the species at a great disadvantage. But he found that this difficulty was overcome by the spores which had been licked from the skin germinating in the gastric juice of the animal's stomach, and, when voided in the excreta, infecting a new host by accidental contact. In the case of both Stereum and Onygena he accomplished for the first time the difficult task of tracing their life-history from spore to fructification.

Ward had prepared himself for the study of bacteria, and in the nineties he undertook, with Prof. Percy Frankland, a prolonged research on behalf of the Royal Society as to the conditions of their occurrence in potable water. The reports of the results fill a thick volume, and the amount of work involved is almost incredible. The bacteriology was entirely due to Ward.

That bacteria are not an inevitable element in potable water is proved by their absence from that of deep springs. They are arrested by filtration through the earth's crust. In any river system they are comparatively fewer towards the watershed, and more frequent towards the mouth. The obvious conclusion is that they are derived from the drainage of the land. As it is known that the bacteria of cholera and typhoid are water-borne, it becomes a problem of vital importance to ascertain if river water is a possible means of distributing these diseases. Ward set to work to ascertain: (i) What was the actual bacterial flora of Thames water; (ii) if this included any pathogenic organisms; (iii) if not, what became of them? The labour required by the first two branches of the enquiry was enormous; he identified and cultivated some eighty species; the resulting answer to the second was happily in the negative.

As to the third, two facts were known. First, that river water, if stored, largely cleared itself of bacteria by mere subsidence; secondly, that Downes and Blunt, in a classical paper communicated to the Royal Society in 1877, had shown that exposure to direct sunlight is fatal to bacteria in a fluid medium.

Ward showed that subsidence could not be entirely relied on, as the sediment might easily become the source of re-infection. The effect of sunlight required more critical examination.

It was known that the spores of anthrax were liable to be washed into rivers. Ward determined to study this as the most extreme type of pathogenic infection. As it is undoubtedly the most deadly micro-organism known, and Ward proposed to deal with it on a large scale, it implied no small degree of courage. He found that the spores of anthrax were effectually killed by a few hours' exposure to even the reflected light of a low winter sun. It was clear that this was due to the direct action of the light and not to any heating effect, apart from the fact that they will tolerate boiling for a few minutes. It was further shown that there was no foundation for the theory of Roux and Duclaux that their death was due to poisoning by products of oxidation of the food-medium. Proof of this, indeed, was hardly required, for Pasteur had shown that the bacteria floating in the atmosphere are mostly dead. Were it not so, no surgical operation would be possible. To the bactericidal effect of sunlight is equally to be attributed the absence of bacteria from the High Alps.

The next point was to ascertain to what rays the effect was due. The spores of anthrax are so minute that, when mixed in large numbers with gelatine, they do not affect its transparency. A plate of glass coated with the mixture is at first clear, but ceases to be so if kept in the dark, owing to the germination of the spores. Ward found, in fact, that a photograph could be printed with it, the darkening being the reverse of that of a silver plate. After experiments with coloured screens he completely solved the problem in 1893, with the aid of apparatus supplied by Sir Oliver Lodge and some advice from Sir Gabriel Stokes, by photographing the spectrum on such a plate. It was at once seen that the destructive effect was due to rays of high refrangibility, and, what was extremely important, extended to, and found its maximum in, the ultra-violet. The same results were obtained with the typhoid bacillus. He made the suggestion that the arc light might be used for the disinfection of hospitals and railway carriages.

Comparatively little was known of the life history of any Schizomycete. Ward therefore made a detailed and exhaustive study of that of *Bacillus ramosus*, the *Wurzel bacillus* of German authors, which is common in Thames water, and bears a superficial resemblance to the anthrax bacillus, but is innocuous. It proved convenient for study, as it ran through its entire life history in from thirty to sixty hours at ordinary temperature. It forms long filaments, the growth of which Ward was able to measure under the microscope with great precision. On plotting out his measurements he obtained a regular curve, from which he found that, under constant conditions, the filament doubled itself in equal times. This he called "the law of doubling." It is the same as the so-called "law of compound interest," and leads to the expression of the growing quantity as an exponential function of the time, so that the time is proportional to the logarithm of that quantity. This relation has, of course, long been familiar in chemical reactions, but, as far as I know, Ward was the first to detect it in any vital process in a plant. This, which was in 1895, has, I think, been overlooked. Stefanowska has since, in 1904, obtained a logarithmic curve for the early period of the growth of maize, which doubles its weight every ten days, and the subject has since been pursued by Chodat and others.

In speculating on the cause of the destructive action of light on bacteria, Ward adopted the view of his friend Elfving, that it inhibited metabolic processes necessary to nutrition. He suggests that the "constructed metabolites" at the moment of assimilation are in a highly unstable condition, and liable to destruction by oxidation promoted by light. He points to the fact that plant structures are frequently provided with colour screens, which would cut off the blue-violet rays and check their action in promoting the rapid oxidation of reserve materials, and he quotes the suggestion of Elfving that chlorophyll itself may serve as such a screen against "destructive metabolic action in synthesis." Ward seems to have attributed little importance to the fact that substantially the same view had long before been put forward by Pringsheim, though received with little favour. His own view that when red and orange predominate in the screens their effect is protective, has since afforded a probable explanation of the colouration of young foliage, especially in the tropics.

It can hardly be doubted that the upshot of Ward's laborious investigations has had a powerful influence in deciding the policy of the future water supply of London. If we hear nothing now of obtaining it from Wales, it is because we know that even polluted flood-water if exposed in large reservoirs will rid itself of its bacterial contamination, partly, as was known already, by subsidence, but most effectually, as shown by Ward, by the destruction of its most deleterious constituents by the direct action of sunlight.

In 1895, Ward was called to the Chair of Botany at Cambridge. He was supported by a distinguished body of fellow-workers, and developed a flourishing school, in which every branch of the science found its scope. The University erected for it an institute which is probably the best equipped in the country, and in March, 1904, I had the pleasure of seeing Ward receive the King and Queen at its inauguration.

During the later years of Ward's life he returned to the study of the Uredineae. The scourge of wheat perhaps from the dawn of agriculture has been "Rust,"

"Ut mala culmos esset rubigo .... intereunt segetes";

and the loss inflicted by it throughout the world is probably not calculable. But the history of the Ceylon coffee disease is only too patent an instance of the injury a uredine can effect.

Eriksson, the most recent authority on the subject, had found himself quite unable to account for sudden outbursts of rust which it did not seem possible to attribute to the result of infection. In 1897 he launched his celebrated theory of the Mycoplasm. He supposed that a cereal subject to rust was permanently diseased and always had been; that the protoplasm of the Uredo-parasite and of the cereal, though discrete, were intermingled and were continuously propagated together; but that while that of the latter was continuously active, that of the former might be latent till called into activity by conditions which favoured it. Ward discussed the theory in his British Association address at Toronto, and was evidently a good deal impressed with it, but nothing short of actual demonstration ever convinced him; and when he proceeded to investigate the actual histological facts on which the theory rested he promptly exploded it.

It is interesting to note that Ward, as I know from correspondence at the time, had himself been embarrassed in investigating the Ceylon coffee disease by the same kind of appearance which had misled Eriksson. It is due to an optical fallacy. When the hypha of a uredine attacks a cell it is unable to perforate it with its whole diameter. It infects it, however, with a reduced and slender filament; this expands again after perforation into a rounded body, the haustorium. In a tangential section the perforating filament cannot be distinguished, and the haustorium looks like an independent body immersed in the cell-protoplasm and with no external connection. It requires a fortunate normal section to reveal what has really taken place. Ward was accordingly able, in a paper in the *Phil. Trans.* in 1903, to dispose conclusively of the mycoplasm. This cleared the ground of an untenable hypothesis. The complicated nature of the problem which still presented itself for investigation can only be briefly indicated. Sir Joseph Banks, whose scientific instinct was sound but curiously inarticulate, had pointed out that the spores entered the stomata, and warned farmers against using rusted litter. Henslow, one of Ward's predecessors in the Cambridge chair, had been confirmed by Tulasne in showing that the uredo- and puccinia-spores (of the barberry) belonged to the same fungus. De Bary traced the germination of the spores and the mode in which the hyphae invaded the host; the fundamental fact, which he observed but did not explain, was that the germinal filament, after growing for a time superficially, bent down to enter the tissues of its host. Pfeffer in 1883 discovered chemotaxis, the directive action of chemical substances on the movement of mobile organisms. De Bary had previously hinted that the hypha might be attracted by some chemical ingredient of the host plant. Myoshi, a pupil of Pfeffer's, showed finally in 1894 that if a plant were injected by a chemotropic substance a fungus-hypha not ordinarily parasitic might be made to behave as such and attack it.

In such circumstances it might seem that the host was not
merely incapable of resisting invasion by the parasite but
actually invited its attack. Nature is, however, not easily baffled
in the struggle for existence. Attack provokes new methods of
defence. Ward soon found himself face to face with "problems
of great complexity," and these occupied the closing years of
his life.

It had been ascertained in fact that the rust fungus is not, as
was at first supposed, a single organism, but comprises, according to Eriksson, thirteen distinct species, each with physiological
varieties, and that those which are destructive to some grasses
and cereals, are incapable of attacking others. This necessitated
a scrutiny of the nature of grass-immunity. In a paper communicated to the Cambridge Philosophical Society in 1902,
Ward announced a conclusion which was as important as it was
unexpected. He had more and more made use of the graphical
method for presenting to the eye at a glance the result of a mass
of separate observations. In this case he uses it with striking
effect. He shows conclusively, as far as rust in brome-grasses
is concerned, that: "The capacity for infection, or for resistance
to infection, is independent of the anatomical structure of the
leaf, and must depend on some other internal factor or factors
in the plant."

Finally, he is led to the conclusion that "it is in the domain
of the invisible biological properties of the living cell that we
must expect the phenomena to reside." He pointed out the
probability that light would be thrown on this from the action
of chemotaxis, on the one hand, and from that of toxins and
antitoxins in animal organisms on the other. This is a most
fertile conception, which would, however, have required a good
deal of verification, and this, unhappily, he did not live to
attempt. But with characteristic ingenuity he pointed out the
analogy between the infective capacity of uredospores and the
prepotency of pollen, which had previously engaged the attention
of Darwin. In a paper published in the following year in the
Berlin *Annales Mycologici*, he announced a no less significant
result. With his usual thoroughness in research he had cultivated
side by side at Cambridge more than two hundred species and

varieties of Bromus, and had watched the degree to which they were infected by rust under identical conditions. He found that though in the brome-grasses the rust peculiar to them is specifically identical its *forms* are highly specialised. The form which attacks the species of one group will not attack those of another. Host and parasite are mutually "attuned." He termed this "adaptive parasitism." This raised the problem, which had first occurred to him in Ceylon, of how a parasite adapted to species of "one circle of alliance" can pass to those of another. Occasionally it happens that a uredo-form will infect a species where it ordinarily fails. In such a case "its uredospore progeny will thenceforth readily infect that species." Ward regarded this as a case of education. Working on this principle, he succeeded by growing the parasite *successively* on a series of allied species which were imperfectly resistant, to ultimately educate it to attack a species hitherto immune. He called these "*bridgeing species.*" He established, in fact, a complete parallelism between the behaviour of rust-fungi and that of pathogenic organisms in animals.

In the midst of this far-reaching research his health began to fail. In 1904 he had been appointed by the Council to represent the Royal Society at the International Congress of Botany held at Vienna in June of the following year. This he attended, though more seriously ill than he was aware of. On his way back he spent three weeks for treatment at Carlsbad, but receiving no benefit, he went, on the advice of Dr Krause, to Dr von Noorden's Klinik at Sachsenhausen (Frankfort). Nothing could be done for him, and he was advised to return home by easy stages. After a period of progressive and extreme weakness, borne with unflinching courage, the end came somewhat suddenly at Torquay on August 26, 1906. He was buried at Cambridge in St Giles's Cemetery on September 3.

From 1880, the year following his degree, Ward never ceased for a quarter of a century to pour out a continuous stream of original work. This alone would be a remarkable performance, had he done nothing else. But he was constantly engaged in teaching work, and he acted as examiner in the Universities of London and Edinburgh. With no less conscientiousness he complied with the demands which the scientific world makes on its members; he served on the Councils of the Royal (1895) and Linnean (1887) Societies; he was President of the Botanical Section of the British Association at Toronto in 1897, and of the Cambridge Philosophical Society in 1904. Beyond all this he found time to give addresses with unfailing freshness of insight; a lecture at the Royal Institution on April 27, 1894, on the "Action of Light on Bacteria and Fungi" was a notable performance; he wrote numerous articles of a more popular kind, and he produced a number of excellent manuals for students on subjects connected with forest, agricultural and pathological botany. Activity so strenuous almost exceeds the limits of human possibility.

Under the influence of Sachs, Ward might have become a distinguished morphologist. But his work in Ceylon led him into a field of research from which he never deviated. A survey of his performance as a whole, such as I have attempted, has a scientific interest of its own. His research was not haphazard. A continuous and developing thread of thought runs through it all. The fundamental problem was the transference of the nutrition of one organism to the service of another. Of this, in Ceylon, Ward found himself confronted with two extreme types, and of both he made an exhaustive study. In Hemileia it was ruthless parasitism; in Strigula advantageous commensalism. Bornet put Schwendener's theory on a firm foundation when he effected the synthesis of a lichen; Ward, in another group, did the same thing for the ginger-beer plant. In such cases the partnership is beneficial. The problem is to trace the process by which one partner gets the upper hand and becomes merely predatory. Ward inherited a strong taste for music, though I believe he never cultivated it. A musical simile may not inappropriately be applied to his work. In its whole it presents itself to me as a symphony in which the education of protoplasm is a recurring *leit-motiv*.

A few words must be said as to his personal characteristics. He had all the qualifications for the kind of research to which he devoted himself. He was singularly dexterous and skilful in manipulation. He was a refined and accomplished draughtsman, and was therefore able to do himself justice by illustration. He was rigorous in demanding exhaustive proof. This almost deteriorated into a defect. He would pursue every side issue which presented itself in a research, and was quite content if it led to nothing. He would say in such a case: "I will not leave a stone unturned." He was apt, too, I think, to attack a problem in too generalised a form. In his nitrogen work it always seemed to me that he wasted energy on remote possibilities, when a clean-cut line of attack would have served him better[1]. But his mind worked in that way, and he could not help himself. It was, I think, one of the most fertile in suggestion that I ever came across. In later years, in conversation especially, thought seemed to come quicker than words to express it. In this respect he reminded one of Lord Kelvin. In such a predicament he would simply remain silent, and slowly move his head. This habit, I think, explains the reputation of being "mysterious" which he seems to have acquired latterly at Cambridge.

He was not without the honour at home which he deserved, apart from the affection of his friends, and had he lived would doubtless have received it from abroad. He was elected F.R.S. in 1888, and received the Royal Medal in 1893. He was elected an Honorary Fellow of Christ's College in 1897, and received an Honorary D.Sc. from the Victoria University in 1902.

Botanical science could ill spare his loss at the early age of 52. But it may be grateful for 25 years of illuminating achievement. It might have been hoped that another quarter of a century would be allotted to one so gifted. But if the "inexorabile fatum" decreed otherwise, he is at least to be numbered amongst those of whom it may be said

"Felix qui potuit rerum cognoscere causas."

[1] Nov. 1911. I must guard myself against the implication that Marshall Ward's method was wrong in principle. For as pointed out by Prof. Turner in his "Address to the Mathematical and Physical Section" of the British Association at Portsmouth the maxim of "leaving no stone unturned" is identical with Prof. Chamberlin's "Method of Multiple Working Hypotheses." And what is at first sight an unlikely hypothesis may turn out to be the true one. Yet the rigorous application of the method is time-consuming and life is short. Some liberty of selection in testing the hypothesis that seems most probable must be allowed the investigator, and the instinct of genius may sometimes hit on the right one.

A SKETCH OF THE PROFESSORS OF BOTANY IN EDINBURGH FROM 1670 UNTIL 1887

By ISAAC BAYLEY BALFOUR

Medicine and Botany—James Sutherland—enforced retirement—the Prestons—Charles Alston—his career—John Hope—Physiological leanings—Daniel Rutherford—Robert Graham—John Hutton Balfour—characteristics—Botanic Society of Edinburgh founded—appointed to Glasgow—transfer to Edinburgh—his numerous activities—laboratory teaching established—field excursions—Ecology—attitude to Darwinism—Alexander Dickson—work in Organography—his versatility.

My task in the warring against oblivion typified in these addresses is to speak about John Hutton Balfour of Edinburgh, one of the botanical teachers of the middle of last century, whose pupils were numbered by thousands, and whose active life bridged the period of the passing of the old and the birth of the new outlook upon science through Darwin's work ; and in relation to what I have to say of him I propose to sketch briefly the stages and development of botanical teaching in Edinburgh from the date when systematised attention was first given to it.

Of the well-recognised fact that the study of Botany as a science has been, to begin with, dependent on Medicine my story furnishes an excellent illustration.

Only towards the end of the seventeenth century had the advance in practice of Medicine in Edinburgh reached a stage which gave urgency to a movement for the improvement in the training of the medical man, and the protection of the public from the attentions of inefficient votaries of the healing art. The foundation of the Royal College of Physicians in 1681 gave expression to the co-operative principle in the control of those who would profess Medicine ; the creation of a Botanic Garden for the purpose of the cultivation of medicinal plants was the response in the direction of safeguarding the practitioner against the herbalist, and of giving him the advantage of a correct knowledge of the plants which were the source of the drugs he himself was to compound. Before this time, whilst many practitioners could grow drug-plants for themselves, and did so, the majority were at the mercy of the herbalist.

Two Edinburgh physicians—(Sir) Robert Sibbald and (Sir) Andrew Balfour—conspicuous among their fellows for their activity in promoting the cause of medical education and in the planning of the Royal College of Physicians, were the pioneers of the study of Botany as a science. Determined that the apprentices in Medicine should have adequate opportunity of learning the sources of many of the drugs in use, they acquired a lease of a small area of land in the neighbourhood of Holyrood Palace in which they arranged to cultivate medicinal plants, stocking it from their own gardens and from those of friends. They secured the services of James Sutherland—described as " knowing " in these matters—and placed their small garden under his care, with the obligation that he should instruct the apprentices and lieges in Botany. Sutherland cultivated his plants so well, and the instruction which he gave was so satisfactory, that ere long—no doubt through Sibbald's influence at Court—a portion of the Royal Flower Garden at Holyrood Palace was assigned for the cultivation of medicinal plants, and thither was transferred the collection already made in the hired area. Thus was founded, with the title of Physick Garden, a Royal Botanic Garden in Scotland, and the first Profession of Botany was set up therein by James Sutherland.

Of the earlier years of Sutherland we have no record. His success as a teacher induced the Town Council of Edinburgh —the body in which was vested at the time all the patronage of the University—to institute a Chair of Botany in the University, and to provide for practical teaching in another Botanic Garden belonging to the town. Sutherland was appointed to the Professorship and also to take charge of this new Town Garden, which, it may interest those who at the present day pass through the Waverley Railway Station to know, occupied a portion of the site of that station. Both these gardens were at some distance from the University, and apparently to save the time of the University students, perhaps also to create a teaching garden entirely within the jurisdiction of the College authorities, another portion of ground occupying a part of the Kirk o' Field, notorious as the place of Darnley's murder, was transformed into a herb-garden. Thus within a few years from the beginning of the movement for the providing of adequate facilities to students for learning about plants, three Botanic Gardens were made available.

During Sutherland's tenure of the Professorship teaching was given by him in these different gardens. It would appear, however, that Sutherland was at heart a numismatist, and whilst during the early period of his incumbency of office he had corresponded with many botanical institutions abroad, had introduced to the gardens new species of plants—many of them now established in the flora—and had published in 1683 a *Catalogue of the plants in the Physical Garden*, in later years his interest was centred in coins and medals. So great was the obsession that the patrons of the University, dissatisfied with his botany, compelled him to resign his Chair in 1706, to which they appointed Charles Preston, but Sutherland retained, until he retired in 1715, charge of the Royal Botanic Garden at Holyrood, of which by Royal Warrant he had been made Keeper with the additional personal recognition of Botanist to the King in Scotland. Thus the increase in number of gardens extended to the Professors, and from 1706 onwards to 1739 there were two rival Botanical Schools in Edinburgh— that of the Royal Garden, and that of the University.

Sutherland's place in relation to the development of scientific Botany in Scotland is that of pioneer in the teaching of systematic Botany from the living plants in relation to Materia Medica, and of first custodian and cultivator of plants for instruction in a public garden. His *Catalogue* is now a book of some rarity—of great rarity in complete state owing to the number of cancel pages—and its reproduction at the present time would have interest alike scientific and historic. It is the first published record of a collection of cultivated plants in Scotland. It tells us the plants which were recognised as indigenous at its date, and from its record we can by correlation with information otherwise obtainable discover the time of introduction to Scotland of alien plants, and thus obtain a basis for gauging their influence on the native Flora as we know it now.

Charles Preston who stepped into the University Chair of Botany vacated in 1706 by Sutherland, was a medical man, an active correspondent of Sloan, Pettiver, and other scientific men in the south. On his death in 1712, after a short tenure of office, George Preston his brother succeeded him and filled the chair until 1739. Both of the Prestons seem to have been chiefly interested in the Materia Medica side of Botany and their teaching was on the lines of it. They are referred to by their contemporaries as men of botanical knowledge and of critical judgment, and their correspondence indicates that they were in touch with the botanical life of their time. Their work in teaching was always in rivalry with that at the Royal Physick Garden. At first no doubt it was effective and useful owing to Sutherland's neglect of his garden, but when a capable active scientific Professor was placed in charge of this Garden the case for such rivalry and duplication of effort ceased, and it is no surprise therefore to find that when a vacancy occurred in 1739 the University Chair was filled by the appointment of the King's Botanist in Charge of the Royal Physick Garden, who was then Dr Charles Alston. And this combination continues to our own time by mutual consent of the Crown and the University.

Sutherland's retirement in 1715 from the Royal Physick Garden four years before his death, which took place in 1719 when he was over 80 years of age, may have been determined by his incapacity for the duties, but it is probable other influences were effective especially as the office of King's

Botanist was a Household Appointment and only during pleasure. Were I merely to tell of incidents in the history of Botany in Edinburgh I would here introduce the story of Dr William Arthur, Sutherland's successor at the Royal Garden. Arthur has no botanical claims, but had influential political friends whose zeal on his behalf he ill requited by becoming one of the leaders in the Jacobite plot to capture the Castle of Edinburgh in 1715. Having failed in the attempt he escaped to Italy, where in 1716 he died from a surfeit of figs! Ignoble fate for a King's Botanist!

A man of real distinction now comes into our botanical history in Charles Alston—a clear observer and experimenter.

Charles Alston, born 24th October, 1685, was the third son of Thomas Alston, M.A. of Edinburgh and M.D. of Caen, one of an old Lanarkshire family settled at Thrinacre Milne and connected with the house of Hamilton. After boyhood at Hamilton, Alston went to the University of Glasgow, but before the period for graduation his father died leaving a widow and large family poorly provided for and young Alston's University career was stopped. Through the intervention of the Duchess of Hamilton Alston was then apprenticed in 1703 to a lawyer with a view to his entering the Estates Office of the Hamilton family. But "anatomy and the shops were more agreeable to him than Style Books or the Parliament House" and his "genius inclined more to Medicine," and in 1709 when the Duchess took him into her service as her "Principal Servant," in which position "he had aboundance of spare time," "he ply'd close the Mathematics and whatever else he thought of use to a student of Medicine, particularly Botany." With this training Alston, through the influence of the Hamilton family, was made King's Botanist, Professor of Botany, and Keeper of the Royal Physick Garden in 1716 after the disappearance of Dr Arthur.

He adopted a wise course on succession. Having put the Garden in such order as he could he hied himself to Leyden in 1718 to study under Boerhaave, and returning thence in August 1719 he graduated in Medicine at the University of Glasgow, became Fellow of the Royal College of Physicians, and in June

1720 was able to begin his botanical lectures in the Garden, followed in November by a course on Materia Medica. These courses he carried on until 1739 when he was given the University Chair of Botany and Materia Medica, and the two Botany Schools were thus merged in one. Alston was now colleague of Munro, Rutherford, Sinclair, and other famous men who at this time were increasing the reputation of the University as a Medical School, and he continued to teach Botany and Materia Medica until his death in 1760.

Alston's teaching was mainly directed to the Materia Medica. His full course of lectures on the subject prepared for publication by himself appeared only as a posthumous work edited by his successor Dr Hope, and they reflect the best knowledge of the time, showing rational scepticism of the efficacy of many simples which experiment had not tested. Essays "On Opium," and "On tin as anthelmintic," and an "Index of Simples" published by him tell of his pharmacological investigations, to which his correspondence with Fothergill and others is also witness. The subject in this line to which he gave most attention and on which he wrote three dissertations based on experiments is that of Quicklime and Water—its efficacy in Calculus and also as an agent for keeping water sweet. From Alston, Stephen Hales, then in touch with the Admiralty upon questions of ventilation and other matters of sanitation, obtained early suggestions, and a long correspondence followed.

Alston, who had to earn his livelihood by medical practice, gave much time to the administration of the Botanic Gardens under his charge, and the elaborate lists which he prepared showing the disposition of plants in the Gardens, witness to his interest in their cultivation. His predilection in systematic arrangement was Tournefortian, and on the promulgation by Linnaeus of his "sexual system" in 1736, no writer was more trenchant than Alston in opposition to it, and by this he became widely known. His criticism was directed against it, not as a method of arranging plants by readily recognised characters, but from the standpoint of denial of the existence of sex. By various experiments as well as by argument, Alston endeavoured to disprove the necessity of the stamens

for the development of fertile seed, citing cases of seed-pro-
duction where no application of the "dust" from the stamens
was possible—thus early recognising conditions which puzzled
botanists for many generations afterwards and until the explanation of apogamy was supplied.   One is tempted to wonder
whether if the Linnaean system had not received the appellation
"sexual" it would have roused the same condemnation from
him as it did.

From his published work, notably the *Dissertation on Botany*
(1754) a translation of a portion of his earlier *Tirocinium
Botanicum Edinburgense* (1740), as also from some MS. of his
lectures which still exist, we recognise the clearness and vigour
of mind of Alston, and the precision of the man is made
abundantly evident in the beautiful copper-plate writing in old
script of his MS.   Page after page is filled without blot or
correction, and the whole systematised and arranged without
flaw.   Anatomical questions were dealt with by him in con-
sonance with the knowledge of the time, mainly resting on
Malpighi; but there is no rational treatment of physiological
subjects, and this is the more surprising inasmuch as he was
in intimate correspondence with Hales, and ought to have been
acquainted with the fundamental experimental work of that
physiologist.   It may be that the fragments of record from
which we have to judge are insufficient for correct appraisement,
but on all the evidence we possess we must conclude that the
two volumes of his *Materia Medica* give us a picture of the
direction of his teaching, and that Botany in the hands of its
leading expositor in Edinburgh was at this period only a hand-
maid to Medicine.

The advent of Alston's successor, John Hope, was the dawn
of new things.   The influence of the work of Hales had
reached Edinburgh.   Comparatively few botanists of to-day
have heard the name of John Hope otherwise than as that of a
correspondent of Linnaeus and ·protagonist in this country of
his system of classification, for these are the claims to distinction assigned to him by the historians of British Botany; and
if one reckons the value of a man's life-work in science by his
published writings alone, that of John Hope would be a minimum;

for only such papers as those " On Rheum palmatum," " On Ferula Assafoetida," "On Eriocaulon septangulare in Scotland," are extant from his pen. Yet John Hope was a botanist inspired by the spirit of research who obtained by scientific experimental work and explained to his pupils facts of plant physiology some of which the botanical world learned from other workers only a hundred years afterwards. It is difficult to account for Hope's reticence. It may be that he intended to give his work to the world in the book upon Botany which had engaged his attention for many years and of which the MS. was in great part ready at the time of his unexpected death in 1786—if so, the botanical world has been the poorer through the want of Hope's book.

But if Hope did not give cause by published contributions to natural knowledge for his recognition in promoting the advance of Botany, he has always been remembered with gratitude for services of administration which he was peculiarly fitted to render and which profoundly affected the study of Botany in Edinburgh.

John Hope was born 10th May, 1725. The son of Robert Hope, a surgeon in Edinburgh, whose father had become one of the Senators of the College of Justice with the title of Lord Rankeillour. Educated at a famous school in Dalkeith, John Hope, who early showed a liking for Botany, entered the University of Edinburgh as a medical student and became a pupil of Alston. His botanical inclinations tempted him to break the course of his medical studies in Edinburgh to study Botany under Bernard de Jussieu in Paris. Returning to Scotland he graduated in Medicine from the University of Glasgow in 1750, joined the Royal College of Physicians in Edinburgh and began medical practice, giving to Botany such time as could be spared from the many ties of a successful practice. In 1760 Alston died, and John Hope became his successor, first of all in 1761 as King's Botanist at Holyrood and subsequently as Professor of Botany and Materia Medica in the University.

Soon after appointment Hope recognised that to continue to hold "colleges" in Materia Medica meant spoliation of his botanical work. The time had come for a separation of the two subjects of Botany and Materia Medica. Problems of the former now pressing were not those specially relating to medicinal plants. He therefore managed to carry through an arrangement by which he retained a chair as Professor of Medicine and Botany, and a new Professorship of Materia Medica was created. The importance of this step for botanical progress was great—it was not merely a question of time occupied but of scientific outlook.

Another movement in the direction of concentration of effort in the cause of Botany was initiated by Hope early in his official career—that for the creation of a new Botanic Garden in a locality outside the immediate influence of town atmosphere, in which the collections distributed over the Holyrood and Town Gardens could be combined. He accomplished his design, and not only this, but obtained from the Crown a permanent endowment for the new Garden. This was no small achievement—but the omens were favourable, for those patrons of science the Earl of Bute and, later, the Duke of Portland, were in power when the Professor made use of the great influence which his family possessed to secure his ends. A spreading city in time made the location of Hope's new Garden unsuitable, and it was transferred to the present site; but it was the effort by Hope which gave the Botanic Garden, and through it Botany, a status among institutions requiring subsidy and maintenance by Government in Scotland, and the obligation so imposed has been upheld notwithstanding an attempt in later years on the part of the Government to get rid of it—an attempt which the short-sighted policy of the University nearly allowed to succeed.

Hope's duties in his University Chair required of him, in addition to his botanical work, clinical teaching in the Hospital, and he also engaged in practice—this for a livelihood—and took active share in the affairs of the Royal College of Physicians, of which he was President at the time of his death, which occurred in 1786. Botany could therefore claim but a portion of his time.

Having established the new Garden, he laboured with assiduity to lay it out effectively, and then to enrich it with

plants. His own ardour and enthusiasm impressed others, and
his pupils in all parts of the world contributed to making the
Garden a renowned collection of the rarest plants. Here Hope
met his students, and here he carried out his many physiological
experiments which gave them instruction.

His teaching was comprehensive. Although no longer tied
by the calls of his Materia Medica, Hope did not ignore the
subject entirely, but plants in this relation were not the groundwork of his instruction. Systematic and descriptive Botany,
recognition of herbs, still found a place in it. In Alston the
most strenuous opponent of the Linnaean method had gone; it
found in Hope a no less strenuous advocate, to whose influence
its rapid adoption in this country owed much. To what extent
Hope made excursions with his pupils, there is no evidence.
His *Hortus Siccus* and lists of plants with localities show that
he was a field-botanist, and in correspondence with, if not more
intimately acquainted with, the botanists who were working out
the Scottish Flora at the period—such men, for instance, as
Lightfoot, Stuart, Robertson. This we do know, that he
encouraged his pupils to investigate the Flora of Scotland,
giving yearly a gold medal for the best Herbarium, and Hope's
"peripatetic pupils" is a designation met with in literature of
the time. This aspect of Hope's teaching, consonant with the
features of the botanical literature of the period, is that which
has been commonly known. It is not however a complete
picture. In Hope Scotland had a physiologist of originality and
skill—who was not only informed upon the work of Hales,
Duhamel, Mariotte and others, but who made his own experiments, clearly devised and effective, and whose catholicity is
attested by his dealing with such problems as growth in length
and thickness, effect of light and gravity, movement of water,
healing of wounds, and the like. This physiology was an
essential element of his teaching, and the effect upon students
of contact with such direct wresting of truth from Nature must
have been immense. Our knowledge of all this, only recently
acquired, throws a new light upon Hope's character, and upon
the influence which he appears to have exercised on the
education of the time. The pity is that he left no published

records, and that this bright period of brilliant research should have become obscured by the scholasticism inherent in the method of classification which he himself did so much to popularise.

In accordance with tradition, the Chair vacated by Hope was filled by the election of another medical practitioner in Edinburgh. Daniel Rutherford was born in Edinburgh 3rd November, 1749, the son of Dr John Rutherford, who as Professor was associated with Alston and others in the reformation of the Edinburgh Medical School. He was distinguished both as a classical scholar and as a mathematician, and after graduating M.A. at the University of Edinburgh, he entered on the medical curriculum, obtaining his diploma of M.D. in 1772. His thesis, when applying for the degree, was " De aero fixo dicto aut Mephitico," and by this he became famous through the distinction he established in it between carbonic acid gas and nitrogen, though he did not give nitrogen its name. The exposition he gave of his precise experimental work has been allowed to entitle him to be regarded as the discoverer of nitrogen, although shortly before the appearance of his thesis Priestley had practically, if less methodically, covered the ground. After graduation, Rutherford travelled in France and Italy, returning to Edinburgh in 1775 to begin the practice of Medicine, becoming Fellow of the Royal College of Physicians, of which he was afterwards President.

Rutherford was a chemist, and I have not discovered in any references to him expressions that would show he was at this period of his life interested in plants otherwise than as objects for his experiments in relation to the chemistry of the atmosphere. In seeking for a reason to explain his selection as Hope's successor in the Chair of Medicine and Botany, one may suggest either the general one of recognition of his scientific ability, or the more special one that in experimenting with plants he had been following on the lines of work so conspicuously developed by Hope. And of course at that time some general knowledge of Botany had to be the possession of every successful physician.

Like his predecessors, Rutherford had to undertake clinical

teaching in the Hospital; he maintained also his private practice, and was keenly interested in the active literary world of his day in which his nephew (Sir) Walter Scott was a brilliant star. The Botanic Garden continued to hold its place as a scientific institution, and from the advent of William McNab as Principal Gardener in 1810, developed into one of the best known in the world. The recording of the plants of Scotland also proceeded apace; two of the Principal Gardeners of the Edinburgh Garden during Rutherford's Keepership—John Mackay from 1800–1802, and George Don from 1802–1806—being foremost in making known its floristic features, and their work Rutherford must have encouraged. From MS. notes of his lectures, I gather that the biological did not attract Rutherford, nor does it appear in the scanty records available that any special development of teaching equipment or of method took place during his tenure of office.

For some years before his death in 1819 Rutherford had been infirm; and speculation as to his successor had been rife. Robert Brown and Sir James Edward Smith were both spoken of. When the vacancy came Robert Brown refused it and Robert Graham, then Professor in the University of Glasgow, was appointed.

Robert Graham was born at Stirling 3rd December, 1786, the third son of Dr Robert Graham of Stirling (afterwards Moir of Leckie). After early education at Stirling, Graham was apprenticed in 1804 to Mr Andrew Wood, Surgeon in Edinburgh, and entered on the study of Medicine at the University, graduating M.D. in 1808. Thereafter he studied at St Bartholomew's Hospital in London for a year before settling in Glasgow, where he was also Lecturer in Clinical Medicine. During this period he published a dissertation "On continued Fever."

Botany in the University of Glasgow at this time had not reached the dignity of having a Professorship. It was attached to the Chair of Anatomy, but a separate lecturer undertook its teaching. To this lectureship Graham was appointed in succession to Dr Brown. This appointment was the prelude to his election as Professor in 1818 when the Chair of Botany was founded—a foundation which owed much to him through his influence with the Duke of Montrose, then Chancellor of the University, of whose house he was a cadet. One of the first efforts of Graham in his new position was directed to the completion of a scheme that was making for the formation of a Botanic Garden. In this he succeeded, and botanical teaching in Glasgow was thus equipped in 1819.

From this sphere in which he had initiated so much, Graham came to Edinburgh in 1820 as Professor of Medicine and Botany and was forced again to take up medical practice and clinical teaching in the Hospital, and in consequence to interest himself in the affairs of the Royal College of Physicians, of which he became President—all this, as in the case of his predecessors, in addition to his botanical work.

His first labour in relation to Botany was to transfer the Botanic Garden which Hope had made to a new site—that which it now occupies. Nearly two years were required to carry out the removal, to the success of which the skill of William McNab, the Principal Gardener, contributed greatly.

During the whole tenure of his offices Graham devoted himself to the affairs of this Garden, and often in the very practical way of supplying funds from his own resources to supplement the inadequate grants obtained from Government. It gave him the material for the description of many new species which were figured in the *Botanical Magazine* and other like periodicals. This systematic botanical work was that which Graham cared for most, it was the backbone of his teaching, and all of his scattered papers deal with this aspect of the subject.

In connection with his teaching Graham developed specially the botanical excursion for the study of Field Botany, making it an integral part of his courses, and in furtherance of its aims travelling far through Scotland—a business of a much more arduous nature in days when railways and motors had not annihilated distance and provided all the comforts of civilisation within easy reach of every district. Graham had intended to publish a *Flora* of Scotland as the result of his practical study of its plants, but it was uncompleted at the time of his death in

1845 after an illness of some duration during which (Sir) Joseph
Dalton Hooker acted as *locum tenens*.

Another new method in his teaching was that of encouraging
students to write essays upon subjects either practical or
theoretical. In this he stimulated investigation. Students in
these days had more time than they have now to devote to such
things, and of their efforts some were sound pieces of research—
the *Botanical Geography* of Hewitt C. Watson first took form in
one of these essays.

John Hutton Balfour[1], who succeeded Graham, was born in
Edinburgh 15th September, 1808. The eldest son of Andrew
Balfour, surgeon in the Army, who afterwards settled in Edinburgh as printer and publisher, in which business his enterprise
was adequate to the venture of the *Edinburgh Encyclopaedia*
under the editorship of (Sir) David Brewster. Andrew Balfour
was a grim old presbyter of the stuff covenanters were made,
and in the strict home environment which he created young
Balfour early came into touch with theological dogma. The echo
of these early impressions remained with him throughout life.

Educated at the High School of Edinburgh where he laid
the foundation of sound classical scholarship—always his
unobtrusive distinction—Balfour entered the curriculum for
the Arts degree at the University. Before completing this he
migrated to St Andrews in order to be under the influence of
Professor Thomas Chalmers—the famous Divine, afterwards
leader in the disruption that founded the Free Church of Scotland—in conformity with the desire of his father that he should
become a minister in the Church of Scotland. But Divinity did
not claim him and he returned to Edinburgh to begin the study
of Medicine—a decision in face of family pressure which is
tribute to the strength of purpose which characterised him and
found expression frequently in after life.

At the beginning of this renewed Edinburgh curriculum
Balfour attended the Botany course of Professor Graham in
1825, and obtained his first scientific instruction in Botany—
a subject for which he had always shown fondness. Robert
Dickson, afterwards Lecturer on Botany at St George's Hospital,

---

[1] His portrait forms the frontispiece of this book.

London, was a fellow-student, and together they, in this and following years, made many botanical excursions about Edinburgh. With his fellows Balfour seems to have been *bon camarade*, acquired all the ephemeral distinction attaching to a facile writer of rhymed couplets for occasions, and as an inveterate maker of puns was in demand for the office of punster at the convivial clubs of the period. A mark of more serious attainment—he was President of the Royal Medical Society in two years. After graduation as M.D., when he also became a fellow of the Royal College of Surgeons in Edinburgh—his thesis for the former being " De Strychnia," for the latter " On Purulent Wounds "—Balfour went in 1832 to Paris to continue his medical education, studying there under Dupuytren, Lisfranc, and Manec. Returning, he settled in Edinburgh in 1834 and entered on practice, becoming assistant within and without the University to Sir George Ballingall, Professor of Military Surgery. Amongst his patients he numbered De Quincey and his family. De Quincey's eldest son died from a cerebral complaint, and the autopsy revealed an interesting pathological condition which formed the subject of Balfour's investigation, and an account of it his first published scientific paper.

From the claims of Medicine Balfour could wrest little time for botanical pursuits, but his holiday always meant the botanical exploration of some area, preferably alpine, and his home became a centre for men of kindred tastes. There in co-operation with his old teacher Graham, and with Greville, Forbes, Falconer, Parnell, Munby and others, was instituted in 1836 the Botanical Society of Edinburgh, with wide aims for the promotion of Botany—amongst them the creation of a botanical library and a herbarium. This has proved a signal service to science. It was the pegging out of a claim which has been made effective. The Society after a life—as with all such societies—of fluctuating periods of greater and lesser activity, flourishes still, and its library and herbarium, transferred to the Crown when the space demand of their bulk became urgent, have been the foundation for the large botanical library and herbarium now maintained and subsidised by Government in the Royal Botanic Garden.

Plants gradually drew Balfour away from patients and in 1840 he carried the divorce so far as to establish himself as a teacher of Botany in the Extra-mural Medical School in Edinburgh—that exemplar of free-trade in teaching—from which so many of the famous occupants of Chairs in the University have entered its portals. But only in 1842, when Sir William Hooker moved to Kew and a vacancy was then caused in the Glasgow Chair of Botany to which Balfour was elected, was he able to give up medical practice entirely.

In Glasgow the first years of Balfour's botanical career were spent, but they were few. On the death of Graham he returned to Edinburgh as Professor of Medicine and Botany and Keeper of the Royal Botanic Garden—the electors passing over Joseph Dalton Hooker also a candidate. In the sphere of these offices the rest of his active life was passed until his retirement in 1879. He came to the University of Edinburgh at a time when the reputation of its medical school was upheld by a remarkable band of teachers in the Medical Faculty—Allen Thomson, Alison, Christison, Goodsir, Gregory, Jameson, Simpson, Syme—and when the struggle of the University after a revised constitution was approaching the climax reached in 1858, when with other Scottish Universities Edinburgh obtained autonomy, and science was enfranchised. Of this Faculty he became Dean, and held office until close upon the time when he became Emeritus. In all the discussions and controversies, destructive and constructive, that attached to so weighty a crisis, Balfour's influence and outlook for science were used with effect, and no less influential were his action and advice in subsequent years when the specific question of medical reform was raised, as it so often was.

Absorbing administrative work of this kind, to which were soon added the duties of a Secretary of the Royal Society of Edinburgh—(and he remained in the Secretariat to the end of his active life)—as well as those of an editor of the *Edinburgh New Philosophical Journal*—(afterwards merged in the *Annals and Magazine of Natural History*)—of Secretary of the Royal Caledonian Horticultural Society and of other offices, made inroad alike upon time and energy of a man who had also the administration of the Royal Botanic Garden in his hands, as well as the calls of his Professorship of Botany to attend to. But Balfour was untiring in industry, prompt and precise in method, and administrative work appealed to him.

Though liable like his predecessors to undertake clinical medical teaching, Balfour, save for occasionally acting as *locum tenens*, took no share in it, and his energies in teaching were devoted to Botany. On the lines he followed he was pioneer. We have seen that Field Botany had been for several decades a characteristic of the Edinburgh Botanic School. Whilst maintaining this feature, Balfour added laboratory work. The word "laboratory" was not then in vogue, and "microscopical room" was the designation of the new domain in which the "guillotine," not the "microtome," was used. In the sphere of practical teaching this was a notable advance, and the more so when the technical difficulties that had to be overcome are remembered— the days of cheap microscopes were but beginning, aniline dyes were not yet. Nevertheless the student of the time had opportunity were he so minded of examining plant-form and plant-structure for himself under direction, and if the equipment for work were not so perfect mechanically as modern methods now permit of, the training in minute observation was no less ex cellent than that of to-day, and the educational effect of the teaching no less valuable. The scheme of work was that of the text-books—passing progressively from tissues to organs vegetative and reproductive both phanerogamic and cryptogamic. The specialisation of the type system had not come.

Before he was able to establish, as he did in the early fifties, practical laboratory classes, Balfour had introduced a system of demonstrations of microscopic objects and of physiological experiments in illustration daily of the subject of his lecture, and it is testimony to his power of infusing zeal in pupils that there was always a contingent of them ready to come to the Botanic Garden at six o'clock in the morning to give voluntary aid in the arranging of these demonstrations for the lecture at eight o'clock. Many of those who came have recorded that they found that period and its work one of the most inspiring in their student history.

This new departure in teaching did not interfere with the continuation and extension of field-work, which up to this time had been the form of practical study cultivated in Edinburgh. On the contrary the Botanical Excursion gave Balfour an outlet for energy and favourable opportunity for the exercise of those gifts of personal magnetism and intellectual stimulus through which he influenced and guided many generations of students. Every Saturday during the summer session an excursion was made, and one of some days' duration usually brought the session to a close.   Through these excursions the greater part of Scotland was traversed—on one occasion the terminal excursion of the session was to Switzerland—and the features of flora and vegetation were brought to the attention of many hundreds of students.

The aim and result of the excursion were not solely the acquisition of plants and their identification.   The stimulating effect on many of this side of Botany is evidenced even in our day by the zeal with which search after rare plants is pursued, and in the eagerness displayed in the race after micro-forms.   But the enticement of acquisition and discovery of novelty whilst there were not the governing influences in Balfour's excursion.   In touch as he was with the problems of organography in its fullest sense, a man of wide reading familiar with the botanical work of his time, and associated as he had been in the field with men like Edward Forbes and Hewett Cottrell Watson, Balfour could and did look at plants from the standpoint of their place in vegetation, and in relation to the conditions of growth, and as having a history in their habitat.   His teaching reflected this. It was never classification, diagnosis, and nomenclature as the end-all of Botany.   The details emphasised changed as the progress of botanical discovery gave new clues to explanation of form and relation, and it was the solvings and attempts at solvings of observed phenomena that gave that fascination to his excursions, the remembrance of which seems to have clung to those who had the fortune to join them.   The succession of plants and plant-form from base to summit of a highland hill ; contrasts of vegetation of stream-course, mountain pasture, alpine rock ; high mountain forms of shore plants ; intrusion and extirpation; factors of distribution and their influence;—those and other problems of what we now term Ecological Botany were themes on which the Professor discoursed in his rambles, filling the pupil with information and forcing him to think out to such conclusion as he might on the evidence before him. And then the whole occasion was so enlivened by the outgo of good humour and mirth in joke and pun and story, that fatigue and weariness, which the physical exercise might evoke in those less attuned than the wiry Professor, were drowned in the sunny current of humanity.

I mention this practical teaching first, for it was the characteristic feature, but the idea of practical illustration pervaded all Balfour's effort. His lecture table became a synopsis of the lecture—living plants, herbarium material, museum specimens, all were pressed into service to elucidate the points of the discourse, whilst the walls were tapestried by diagrams. Never did teacher more sedulously absorb the new for presentation to his pupils. He was a lucid expositor, and, apart from his University lectures, during many years was sought after for more popular discourses to non-academic audiences.

The period of Balfour's teaching included the momentous year 1860. The impulse of the new spirit introduced by Darwin did not stimulate Balfour as it might have done a younger man. His religious beliefs—always in evidence—were showing then the influence of his early environment, and whilst Darwin's work was incorporated in his teaching, the acceptance of Darwin's theory appeared too near the negation of faith. On Balfour indeed, as on others with like views, the immediate effect of the *Origin* was the opposite of vivifying. It gave a shock. And this, I conceive, not so much a consequence of Darwin's own statement of his theory as of the forceful uncompromising attitude of the chief protagonist of his cause. Arrogance there was on the religious side, but no less also on the scientific side in the discussion. Perhaps it was well that the contest was sharp and bitter. It ended sooner, but its course was strewn with misconceptions and with confusion of cause and effect. In our days of complete reconciliation, when every tyro lisps in phyletic numbers as the outcome of Darwin's work, it is not amiss to recall the struggle at its inception—lest we forget.

The system of Essays which formed so important a part of Graham's teaching remained as prominent and was even developed further in Balfour's course in a way which had the inestimable merit of making the student feel that his study of plants had a living relationship with the everyday concerns of life. Thus when Simpson was engaged in his epoch-making investigations on anaesthetics, the subject for an essay was the effect of anaesthetics on sensitive plants, and by way of emphasis, the prize awarded was a gift by Simpson himself. Similarly Balfour enlisted the sympathy of Messrs Lawson, the prominent agricultural nurserymen of the day, and their prizes for dissection of grasses, for kinds of cereals, and like subjects, were constant reminders of the relations of botanical study to agriculture. The subjects of essays covered a wide field. The titles— influence of narcotic and irritant gases, changes which have taken place in the Flora of Britain during the historical era, cytogenesis and cell development, phanerogamous embryology, cryptogamous reproduction, teratology—may serve to indicate this, and an essential was always the practical illustration, microscopic or other.

For the use of the students Balfour compiled text-books which, like his lectures, are comprehensive in the field they cover, and encyclopaedic in the information they convey. His facile pen found expression too in numberless articles in encyclopaedias and magazines, and his activity as an expositor of botanical topics of the time was unbounded.

In the Botanic Garden Balfour obtained the material for the definite contributions he made to natural knowledge which are in the domain of Systematic Botany. No work in which Balfour engaged gave him more genuine pleasure than the administration of the Botanic Garden. Entering on the responsibility of its care when its repute was high, he left it on laying down office in even higher reputation, for in the McNabs—William and James—father and son—he had lieutenants of the first rank in gardening. During his regime the equipment for laboratory teaching to which reference has been made was installed, a museum to which old pupils all over the world contributed was instituted, and the Garden itself trebled in size, the latest addition, made just before his retirement, being an area to be cultivated as an arboretum for students of Forestry—a subject then beginning to claim attention.

With Balfour's retirement in 1879 the link of Botany with Medicine in the University was still further weakened. Medicine was left out of the title of the Chair to which Alexander Dickson succeeded.

Alexander Dickson of Hartree and Kilbucho was born at Edinburgh, 21st July, 1836. He was the second son of David Dickson of Hartree in Peeblesshire, and the representative of a family for long lairds of the estates of which, by the early death of his elder brother, he became proprietor. Educated privately, he entered the University of Edinburgh as a student of Medicine, graduating in 1860. Before graduation he had studied in Würzburg and in Berlin, particularly under Kölliker and Virchow, and after it he embarked on the stream of medical practice in Edinburgh. But that was convention—a demonstration of brass plate. His means placed him beyond the necessity of such professional work. His instinct lay in the direction of discovery of method more than in its application. During his student days he had shown a keen interest in Botany. Before graduation he had written on botanical subjects, and his thesis on graduation "The development of the flower in Caryophyllaceae" witnesses to his obsession. Whilst waiting for patients, he had continued work on embryogeny in plants, and when in 1862 the ill health of Professor Dickie at Aberdeen required the appointment of a substitute, the selection of Dickson set seal to his claims as a professed Botanist. In 1866 he succeeded Harvey as Professor in Dublin. Thence in 1868 he was translated to Glasgow as successor to Walker Arnott, and in 1879 became Professor of Botany and Queen's Botanist in Edinburgh on the retirement of Balfour, and, holding these positions, he died in 1887.

Dickson's passion was not teaching, and his success is testimony to the quality of the man. He was adored by his students, as could not well be otherwise with a man of his geniality and kindliness; he took immense pains over his lectures, spending hours daily over the making of fresh drawings on the blackboard for his classes, holding that a student would copy a temporary sketch although he would not copy a permanent wall-diagram; the lecture itself was a model of scientific presentment; at excursions he was untiring in demonstration and in fruitful suggestion, and he was always ready to give of his best to his pupils; but his real love was for research and he carried out many organographical investigations which have added to the sum of natural knowledge. His record in published papers far exceeds that of any of his predecessors, and the quality of his work recalls that of Irmisch. Flower-morphology, embryogeny, teratology, were the subjects to which he gave most attention in research, and in them he obtained results of solid and permanent value. For a time the subject of phyllotaxy occupied him, but it is not a fruitful theme although it gave him opportunity for showing his power of clear analysis; much more interesting was his subsequent work on pitcher plants of kinds.

Dickson possessed great skill in manipulation, and was strikingly effective in the use of his pencil in artistic delineation of the objects of his investigation. Careful in his work he took endless pains to secure that accuracy which it always shows. Further, his subject is always illumined by the comparative method of treatment which his wide knowledge and sound critical faculty enabled him to bring to bear upon it.

The duties of his lairdship were no light ones to Dickson who had set himself to build up again what had come to him in an impoverished condition, and affairs of Church and State were a very real interest to him. Amidst all these ties, to which has to be added the administration of the Botanic Garden, in which during his tenure a new and enlarged Lecture Hall was built, he found time to cultivate the musical faculty for which he was distinguished; not only was he a pianist of mark, but he found absorbing zest in the collecting of national airs sung by the peasants of Scotland.

In the line of Professors of Botany in Edinburgh no one ranked higher in distinction than Alexander Dickson, with whose name I conclude this sketch.

SIR JOSEPH DALTON HOOKER

1817—1911

BY F. O. BOWER.

His long life — childhood and education — travels — Geological work — Morphological Memoirs—administrative duties—systematic works— relations with Darwin — acceptance of Mutability of Species — his philosophical Essays—their influence in advancing Evolutionary Belief.

IT is a difficult task to condense within suitable limits an appreciation of so long and strenuous a life as that of Sir Joseph Hooker. Naturally with age the bodily strength waned, but the vivid mind remained unimpaired to the end. He even continued his detailed observations till very shortly before his death in December, 1911. The list of his published works extends from 1837 to 1911, a record hardly to be equalled in any walk of intellectual life.

Sir Joseph Hooker was born at Halesworth, in Suffolk, in 1817. His father, Sir William Hooker, brought him to Glasgow as a child of four years of age, when he entered on his duties as Professor of Botany in 1821. The Professor established himself in Woodside Crescent, conveniently near to the Botanic Garden, then but recently established, but developing under his hands with wonderful rapidity. Doubtless his little son was familiar with it and its contents from childhood. He grew up in an atmosphere surcharged with the very science he was to do so much to advance. His father's home was the scene of manifold activities. It housed a rapidly growing herbarium and museum. It was there that the drawings were made to illustrate that amazing stream of descriptive works which Sir William was then producing. New species must have been almost daily

Plate XXVI

SIR JOSEPH DALTON HOOKER
(From the photograph by Mrs Cameron, 1868)

under examination, often as living specimens. Between the garden and the house the boy must have witnessed constantly, during the most receptive years of childhood, the working of an establishment that was at the time without its equal in this country, or probably in any other. The eye and the memory must have been trained almost unconsciously. A knowledge of plants would be acquired as a natural consequence of the surroundings, and without the effort entailed by study in later years. Few ever have known, or ever will know, plants as he did. Such knowledge comes only from growing up with them from earliest childhood.

Side by side with this almost unconscious education in Botany the ordinary curriculum of school and of college was pursued. There is no record of academic successes either at the High School, or at the University of Glasgow, beyond a prize "for the best Essay on the Brain and Nerves," in 1836. But the following year saw his first publication : for he described, while still a student, three new species of Mosses. It may be remarked that, like his father, his first writings related to the lower Plants. He never lost his interest in them, though in later years duty diverted him to the study of the Flowering Plants. An incident of his student period, which he himself relates, is, however, a more clear indication of the life that was to follow than any early publication of new species. He tells how an opportunity was given him of reading the proofs of Darwin's *Voyage of the Beagle.* " I was hurrying on my studies (that is for the final examination in Medicine)...and so pressed for time was I that I used to sleep with the sheets of ' The Journal' under my pillow, that I might read them between waking and rising. They impressed me profoundly, whilst they stimulated me to enthusiasm in the desire to travel and observe." The opportunity came to him almost at once in the four years' voyage to the Antarctic. At the age of 22, having passed his examinations, and graduated as M.D., he was equipped at every necessary point for his duties as Assistant Surgeon and Botanist in the " Erebus," then about to start, along with the " Terror," on the famous voyage under the command of Sir James Clark Ross.

No attempt will here be made to give any consecutive

biographical sketch of Sir Joseph Hooker. Several such have already appeared. The interest of the reader will be more readily engaged by indicating the various lines of activity in which he excelled. He was never a professional teacher, except for a short period of service as deputy for Graham in Edinburgh. There was a moment when he might have been Professor in Edinburgh, but it passed. He left no pupils, except in the sense that all botanists have learned from him through his books. We shall contemplate him rather as a Traveller and Geographer, as a Geologist, as a Morphologist, as an Administrator, as a Scientific Systematist, and above all as a Philosophical Biologist. He played each of these several parts in the Drama of Science. The endeavour will be made, however imperfectly, to touch upon them all.

The experiences of Hooker as a *traveller* began immediately after taking his degree, with his commission in 1839 as Assistant Surgeon and Botanist in the "Erebus." Scientific Exploration was still in its heroic age. Darwin was only three years back from the voyage of the "Beagle." We may well hold the years from 1831, when the "Beagle" sailed, to 1851, when Hooker returned from his Indian journey, or 1852, when Wallace returned from the Amazon, to have been its golden period. Certainly it was if we measure by results. Unmatched opportunity for travel in remote and unknown lands was then combined with unmatched capacity of those who engaged in it. Nor was this a mere matter of chance. For Darwin, Wallace, and Hooker all seized, if they did not in some measure make, their opportunity.

The intrepid Ross, with his two sailing ships, the "Erebus" and the "Terror," probed at suitable seasons during four years the extreme south. The very names of the Great Ice Barrier, M'Murdo Sound, Mount Erebus and Mount Terror, made familiar to us by adventures seventy years later under steam, remain to mark some of his additions to the map of the world. Young Hooker took his full share of risks, up to the point of being peremptorily ordered back on one occasion by his commanding officer. To his activity and willingness, combined with an opportunity that can never recur in the same form, is due that great collection of specimens, and that wide body of fact which he acquired. On the outward and return voyages, or in the intervals when the season was not favourable for entering the extreme southern seas, the expedition visited Ascension, St Helena, the Cape, New Zealand, Australia, Tasmania, Kerguelen Island, Tierra del Fuego, and the Falkland Islands. The prime object of the voyage was a magnetic survey, and this determined its course. But it brought this secondary consequence; that Hooker had the chance of observing and collecting upon all the great circumpolar areas of the southern hemisphere. The results he later welded together into his first great work, *The Antarctic Flora.*

Very soon after his return from the Antarctic the craving for travel broke out afresh in him. He longed to see a tropical Flora in a mountainous country, and to compare it at different levels with that of temperate and arctic zones. Two alternatives arose before him : the Andes and the Himalaya. He chose the latter, being influenced by promises of assistance from Dr Falconer, the Superintendent of the Calcutta Garden. But before he left England his journey came under the recognition of Government. He not only received grants on the condition that the collections made should be located in the Herbarium at Kew, but he was accredited by the Indian Government to the Rulers, and the British Residents, in the countries whose hitherto untrodden ways he was to explore. After passing the cold season of 1848 in making himself acquainted with the vegetation of the plains and hills of Western Bengal, he struck north to the Sikkim Himalaya. Hither he had been directed by Lord Auckland and by Dr Falconer, as to ground unbroken by traveller or naturalist. The story of this remarkable journey, its results and its vicissitudes, including the forcible detention of himself and his companion Dr Campbell by a faction of the Court of Sikkim, is to be found in his *Himalayan Journals.* These most fascinating volumes of travel were published in 1854. They tell how he spent two years in the botanical exploration and topographical survey of the state of Sikkim, and of a number of the passes leading into Thibet ; and how towards the close of 1848 he even crossed the western frontier of Sikkim, and explored a portion of Nepal that has never since been open to travellers. In 1849 he returned to Darjeeling, and busied himself with arranging his vast collections. Here he was joined by an old fellow-student of Glasgow, Dr Thomas Thomson, son of the professor of that name. The two friends spent the year 1850 in the botanical investigation of Eastern Bengal, Chittagong, Silhet, and the Khasia hills. In 1851 they returned together to England.

The botanical results of these Indian journeys were immense, and they provided the material for much of Hooker's later scientific writing. Nearly 7000 species of Indian plants were collected by these two Glasgow graduates. But Hooker was not a mere specialist. His *Journals* are full of other observations, ethnographical, ornithological, and entomological. His topographical results especially were of the highest importance. They formed the basis of a map published by the Indian Topographical Survey. By the aid of it the operations of various campaigns and political missions have since been carried to a successful issue. If he were not known as a Botanist, he would still have his assured place as a Geographer.

After his return from India, nine years ensued of quiet work at home. But in 1860 Hooker took part in a scientific visit to Syria and Palestine, ascending Mount Lebanon, where he specially paid attention to the decadent condition of the Cedars, his observations leading later to a general discussion of the genus. Again a period of ten years intervened, his next objective being Morocco. In 1871, with Mr Ball and Mr Maw, he penetrated the Atlas Range, never before examined botanically. His last great journey was in 1877, when he was sixty years of age. With his old friend, Prof. Asa Gray of Harvard, he visited Colorado, Wyoming, Utah, the Rocky Mountains, the Sierra Nevada, and California. Prof. Coulter of Chicago, who was one of the party in the Rockies, has told me how difficult it was to round up the two elderly enthusiasts to camp at night.

This is an extraordinary record of travel, especially so when we remember that all the journeys were fitted into the intervals of an otherwise busy life of scientific work and administration. At one time or another he had touched upon every great continental area of the earth's surface. Many isolated islands had also been examined by him, especially on the Antarctic voyage. Not only were fresh regions thus opened up for survey and collection, but each objective of the later journeys was definitely chosen for scientific reasons. Each expedition helped to suggest or to solve major problems. Such problems related not only to the distribution, but also to the very origin of species. Darwin saw this with unerring judgment as early as 1845. Hooker was then but twenty-eight years old, and the records of the Antarctic voyage were only in preparation. Nevertheless Darwin wrote with full assurance in a letter to Hooker himself: "I know I shall live to see you the first authority in Europe on that grand subject, that almost keystone of the laws of Creation, Geographical Distribution." Never was a forecast more fully justified. But that position, which Hooker undoubtedly had, could only have been attained through his personal experience as a traveller. Observation at first hand was the foundation upon which he chiefly worked. Hooker the traveller prepared the way for Hooker the philosopher.

Sir Joseph Hooker would probably have declined to consider himself as a *Geologist*. He was, however, for some eighteen months official Botanist to the Geological Survey of Great Britain. He was appointed in April 1846, but relinquished the post in November 1847 in order to start on his Himalayan journey. During that short period three Memoirs were published by him on Plants of the Coal Period. They embodied results derived from the microscopic examination of plant-tissues preserved in Coal Balls, a study then newly introduced by Witham, and advanced by Mr Binney. It has since been greatly developed in this country. Such studies were continued by him at intervals up to 1855. While he was thus among the first to engage in this branch of enquiry, he may be said to have originated another line of study, since largely pursued by geologists. For he examined samples of diatomaceous ooze from the ocean-floor of the Antarctic, and so initiated the systematic treatment of the organic deposits of the deep sea. Yet another branch of geological enquiry was advanced by him in the Himalaya. For there he made observations on the glaciers of that great mountain chain, his notes supplying valuable material to both Lyell and Darwin. He also accumulated valuable data concerning the stupendous effects of sub-aerial denudation at great elevations. His latest contribution of a geological character was in 1889, when he returned to an old problem of his youth, the Silurian fossil *Pachytheca*. But he had to leave the question of its nature still unsolved. This geological record is not an extensive one. But the quality and rapidity of the work showed that it was the time and opportunity and not the faculties that were wanting. Moreover, it is worthy of remark that the problems he handled were all nascent at the time he worked upon them.

The list of Sir Joseph Hooker's memoirs which deal *morphologically* with more limited subjects than is possible in floristic works, is a restricted one. In 1856 he produced a monograph on the Balanophoraceae, based upon collections of material from the most varied sources. It is still an authority very widely quoted on these strange parasites. In 1859 he described the development and structure of the Pitchers of *Nepenthes*, while the physiological significance of these, and other organs of carnivorous plants, formed the subject of an Address before the British Association at Belfast, in 1874. And in 1863 his great monograph appeared upon that most remarkable of all Gymnospermic plants, *Welwitschia*. These works bore the character of a later period than the time when they were produced. In Britain, between 1840 and 1875, investigation in the laboratory, by microscopic analysis of tissues, was almost throttled by the overwhelming success of systematic and descriptive work. The revival of investigation in the laboratory rather than that in the herbarium dates from about 1875. But we see that Hooker was one of the few who, prior to that revival, pursued careful microscopic analysis side by side with systematic and floristic work.

The noble establishment of the Royal Gardens at Kew is often spoken of as the Mecca of Botanists. It is also the Paradise of the populace of London. It was the Hookers, father and son, who made Kew what it is. When we contemplate Sir Joseph as an *administrator*, we immediately think of the great establishment which he and his father ruled during the first half century of its history as a public institution. Kew had existed for long as a Royal Appanage before it was handed over to the Nation. The Botanic Garden had, indeed, ranked for upwards of half a century as the richest in the world. But after the death of King George III. it had retrograded scientifically. On the accession of Queen Victoria a revision of the Royal Household had become necessary. It was then decided to transfer the garden to the Commissioners of Woods and Forests. This took place in 1840, and in 1841 Sir William Hooker, who was then Professor in Glasgow, was appointed the first Director. The move to Kew, whither he took his private Library, Herbarium, and Museum, was carried out in the absence of his son, who was still in the Antarctic. It was not till the Himalayan journey was over in 1851 that Sir Joseph settled at Kew, his great collections having already been consigned there by agreement with the Government. In 1855 he was appointed assistant to his father in the Directorship. Finally, he became himself Director on his father's death in 1865, and he held the position for twenty years.

So long associated together, it is difficult to disentangle the parts that father and son actually played in the creation of Kew as it now is. Nor is there need to attempt it. The original area of the Garden at Kew was less than 20 acres. But in 1855, when Sir Joseph joined his father in the directorate, it had grown by successive additions to 70 acres. Finally, the large area of 650 acres came under the Director's control. Numerous large glass houses were built. Three Museums were established, and the vast Herbarium and Library founded and developed. The Garden Staff rose to more than 100 men. The day-by-day administration of such an establishment would necessarily make great demands upon the time, energy, tact, and skill of its official head. But in addition there was the growing correspondence to be attended to, on the one hand with botanists all over the world, on the other with the Government Departments, and especially with the Indian and Colonial Offices. As the activity of the Garden extended, there grew up a large staff of scientific experts and artists, whose duties centred round the Herbarium and Library. These all looked to the Director for

their guidance and control. The descriptive work prepared by
them for publication took formidable dimensions. The production of the Floras of India, and of the Colonies, the publication
of which was conducted under Government subvention, had to
be organised and carried through. These matters are mentioned
here so as to give some idea of the extent and complexity of the
work which was being carried on at Kew. For ten years as
Assistant Director, and for twenty years as Director, Sir Joseph
Hooker guided this complex machine. The efficiency of his
rule was shown by the increasing estimation in which the Garden
was held by all who were able to judge.

It was the founding of the Herbarium and Library at Kew
which, more than anything else, strengthened the scientific
establishment. As taken over from the Crown the Garden
possessed neither. But Sir William brought with him from
Glasgow his own collections, already the most extensive in
private hands. For long years after coming to Kew he maintained and added to his store at his own expense. But finally
his collections were acquired after his death by Government.
His Herbarium was merged with the fine Herbarium of
Bentham, already presented to the nation in 1857. Thus, the
opening years of Sir Joseph's directorate saw the organisation
upon a public basis of that magnificent Herbarium and Library,
which now contains not only his father's collections, but also
his own. Among the enormous additions since made to the
Herbarium of Kew, its greatest interest will always be centred
in the Hookerian collections which it contains.

It might be thought that such drafts as these upon the time
and energies of a scientific man would leave no opportunity for
other duties. But it was while burdened with the directorship
that Sir Joseph was called to the highest administrative office in
science in Great Britain. He served as President of the Royal
Society from 1873 to 1878. The obligations of that position are
far from being limited to the requirements of the Society itself.
The Government of the day has always been in the habit of
taking its president and officials into consultation in scientific
matters of public importance. In these years the administrative demands upon Sir Joseph were the greatest of his life

They are marked by a temporary pause in the stream of publication. None of his own larger works belong to this period. It happens only too often in this country that our ablest men are thus paralysed in their scientific careers by the potent vortex of administration. Not a few succumb, and cease altogether to produce. They are caught as in the eddy of the Lorelei, and are so hopelessly entangled that they never emerge again. They fail to realise, or realise too late, that the administration of matters relating to a science is not an end in itself, but only a means to an end. Some, the steadfast and invincible seekers after truth, though held by the eddy for a time, pass again into the main stream. Hooker was one of these. The Presidency of the Royal Society ended at the usual term of five years. Seven years later he demitted office as Director of Kew. He was thus free in 1885, still a young man in vigour though not in years. For over a quarter of a century after retirement he devoted the energy of his old age to peculiarly fruitful scientific work. Thus the administrative tie upon him was only temporary. So long as it lasted he faithfully obeyed the call of duty, notwithstanding the restrictions it imposed.

No exhaustive catalogue need be given of the works upon which the reputation of Sir Joseph Hooker as a *scientific systematist* was founded. It must suffice briefly to consider his four greatest systematic works, *The Antarctic Flora*, *The Flora of British India*, *The Genera Plantarum*, and the *Index Kewensis*.

We have seen how on the Antarctic voyage Hooker had the opportunity of collecting on all the great circumpolar areas of the Southern Hemisphere. His *Antarctic Flora* was based on the collections and observations then made. It was published in six large quarto volumes. The first related to the Lord Auckland and Campbell Islands (1843—1845); the second to Fuegia and the Falkland Islands (1845—1847); the third and fourth to New Zealand (1851—1853); and the fifth and sixth to Tasmania (1853—1860). They describe about 3000 species, while on 530 plates 1095 species are depicted, usually with detailed analytical drawings. But these volumes did not merely contain reports of explorations, or descriptions of the many new species collected. There is much more than this in them. All the known facts that could be gathered were incorporated, so that they became systematically elaborated and complete Floras of the several countries. Moreover, in the last of them, the *Flora Tasmaniae*, there is an Introductory Essay, which in itself would have made Hooker famous. We shall return to this later. Meanwhile we recognise that the publication of the *Botanical Results of Ross's Voyage* established Hooker's reputation as a Traveller and Botanist of the first rank.

What he did for the Antarctic in his youth he continued in mature life for British India. While the publication of the *Antarctic Flora* was still in progress, he made his Indian journeys. The vast collections amassed by himself and Dr Thomson were consigned by agreement with Government to Kew. Thither had also been brought in 1858 "seven waggon-loads of collections from the cellars of the India House in Leadenhall Street, where they had been accumulating for many years." They included the herbaria of Falconer and Griffith. Such materials, with other large additions made from time to time, flowed into the already rich Herbarium at Kew. This was the material upon which Sir Joseph Hooker was to base his *Magnum Opus*, the *Flora of British India*.

Already in 1855 Sir Joseph, with his Glasgow college friend, Thomas Thomson, had essayed to prepare a " Flora Indica." It never advanced beyond its first volume. But if it had been completed on the scale set by that volume, it would have reached nearly 12,000 pages! After a pause of over fifteen years Hooker made a fresh start, aided now by a staff of collaborators, and the *Flora of British India* was the result. It was conceived, he says with regret, upon a restricted plan. Nevertheless it ran to seven volumes, published between the years 1872 and 1897. There are nearly 6000 pages of letterpress, relating to 16,000 species. It is, he says in the Preface, a pioneer work, and necessarily incomplete. But he hopes it may "help the phytographer to discuss problems of distribution of plants from the point of view of what is perhaps the richest, and is certainly the most varied botanical area on the surface of the globe."

Scarcely was this great work ended when Dr Trimen died. He left the *Ceylon Flora*, on which he had been engaged, incomplete.

Three volumes were already published, but the fourth was far from finished, and the fifth hardly touched. The Ceylon Government applied to Hooker, and though he was now eighty years of age, he responded to the call. The completing volumes were issued in 1898 and 1900. This was no mere raking over afresh the materials worked already into the *Indian Flora*. For Ceylon includes a strong Malayan element in its vegetation. It has, moreover, a very large number of endemic species, and even genera. This last floristic work of Sir Joseph may be held fitly to round off his treatment of the Indian Peninsula. His last contribution to its botany was in the form of a "Sketch of the Vegetation of the Indian Empire," including Ceylon, Burma, and the Malay Peninsula. It was written for the *Imperial Gazetteer*, at the request of the Government of India. No one could have been so well qualified for this as the veteran who had spent more than half a century in preparation for it. It was published in 1904, and forms the natural close to the most remarkable study of a vast and varied Flora that has ever been carried through by one ruling mind.

The third of the systematic works selected for our consideration is the *Genera Plantarum*. It was produced in collaboration with Mr Bentham. Of its three massive volumes the first was published in 1865, and the work was completed in 1883. It consists of a codification of the Latin diagnoses of all the genera of Flowering Plants. It is essentially a work for the technical botanist, but for him it is indispensable. Of the known species of plants many show such close similarity of their characters that their kinship is recognised by grouping them into genera. In order that these genera may be accurately defined it is necessary to have a *précis* of the characters which their species have in common. This must be so drawn that it shall also serve for purposes of diagnosis from allied genera. Such drafting requires not only a keen appreciation of fact, but also the verbal clearness and accuracy of the conveyancing barrister. The facts could only be obtained by access to a reliable and rich Herbarium. Bentham and Hooker, working together at Kew, satisfied these drastic requirements more fully than any botanists of their time. The only real predecessors of this monumental

work were the *Genera Plantarum* of Linnaeus (1737—1764) and
of Jussieu (1789), to which may be added that of Endlicher
(1836—1840). But all of these were written while the number
of known genera and species was smaller. The difficulty of the
task of Bentham and Hooker was greatly enhanced by their
wider knowledge. But their *Genera Plantarum* is on that
account a nearer approach to finality. Hitherto its supremacy
has not been challenged.

The fourth of the great systematic works of Hooker mentioned above was the *Index Kewensis*. It was produced upon
the plan and under the supervision of Sir Joseph by Dr Daydon
Jackson and a staff of clerks. The publication began in 1893,
and successive supplements to its four quarto volumes are still
appearing at intervals. The expense was borne by Charles
Darwin. The scheme originated in the difficulty he had found
in the accurate naming of plants. For "synonyms" have frequently been given by different writers to the same species, and
this had led to endless confusion. The object of the *Index* was
to provide an authoritative list of all the names that have been
used, with reference to the author of each and to its place of
publication. The habitat of the plant was also to be given.
The correct name in use according to certain well-recognised
rules of nomenclature was to be indicated by type different from
that of the synonyms superseded by it. The only predecessor
of such an Index was Steudel's *Nomenclator Botanicus*, a book
greatly prized by Darwin, though long out of date. He wished
at first to produce a modern edition of Steudel's *Nomenclator*
This idea was, however, amended, and it was resolved to
construct a new list of genera and species, founded upon Bentham
and Hooker's *Genera Plantarum*. Sir Joseph Hooker was asked
by Mr Darwin to take into consideration the extent and scope
of the proposed work, and to suggest the best means of having it
executed. He undertook the task, and it was he who laid out
the lines to be followed. After years of labour by Dr Daydon
Jackson and his staff, the work was produced. But Sir Joseph
read and narrowly criticised all the proofs. Imagine four large
quarto volumes, containing in the aggregate 2500 pages, each
page bearing three columns of close print, and each column

about fifty names. The total figures out to about 375,000 specific names, all of which were critically considered by the octogenarian editor! Surely no greater technical benefit was ever conferred upon a future generation by the veterans of science than this *Index*. It smooths the way for every systematist who comes after. It stands as a monument to an intimate friendship. It bears witness to the munificence of Darwin, and the ungrudging personal care of Hooker.

But the author of great works such as these was still willing to help those of less ambitious flights. I must not omit to mention two books which, being more modest in their scope, have reached the hands of many in this country. In 1870 Hooker produced his *Students' Flora of the British Islands*, of which later editions appeared in 1878 and 1884. It was published in order to "supply students and field botanists with a fuller account of the plants of the British Isles than the manuals hitherto in use aim at giving." In 1887 he edited, after the death of its author, the fifth edition of Bentham's *Handbook of the British Flora*. Both of these still hold the field, though they require to be brought up to date in point of classification and nomenclature.

The object of these brief sketches of four of the great systematic works of Sir Joseph Hooker has been to show how fully he was imbued with the old systematic methods : how he advanced, improved and extended them, and was in his time their chief exponent. His father had held a similar position in the generation before him. But the elder Hooker, true to his generation, treated his species as fixed and immutable. He did not generalise from them. His end was attained by their accurate recognition, delineation, description, and classification. The younger Hooker, while in this work he was not a whit behind the best of his predecessors, saw further than they. He was not satisfied with the mere record of species as they were. He sought to penetrate the mystery of the origin of species. In fact, he was not merely a Scientific Systematist in the older sense. He was a *Philosophical Biologist* in the new and nascent sense of the middle period of the nineteenth century. He was an almost life-long friend of Charles Darwin. He was the first confidant of his species theory, and, excepting Wallace, its first

whole-hearted adherent.   But he was also Darwin's constant
and welcome adviser and critic.   Well indeed was it for the
successful launch of evolutionary theory that old-fashioned
systematists took it in hand.   Both Darwin and Hooker had
wide and detailed knowledge of species as the starting-point of
their induction.

Before we trace the part which Hooker himself played in the
drama of evolutionary theory, it will be well to glance at his
personal relations with Darwin himself.   It has been seen how
he read the proof-sheets of the *Voyage of the 'Beagle'* while still
in his last year of medical study.   But before he started for the
Antarctic he was introduced to its author.   It was in Trafalgar
Square, and the interview was brief but cordial.   On returning
from the Antarctic, correspondence was opened in 1843.   In
January 1844 Hooker received the memorable letter confiding to
him the germ of the Theory of Descent.   Darwin wrote thus:
" At last gleams of light have come, and I am almost convinced
that species are not (it is like confessing a murder) immutable:—
I think I have found (here's presumption!) the simple way by
which species become exquisitely adapted to various ends."
This was probably the first communication by Darwin of his
species-theory to any scientific colleague.

The correspondence thus happily initiated between Darwin
and Hooker is preserved in the *Life and Letters of Charles
Darwin*, and in the two volumes of *Letters* subsequently published.   They show on the one hand the rapid growth of a deep
friendship between these two potent minds, which ended only
beside the grave of Darwin in Westminster Abbey.   But what
is more important is that these letters reveal, in a way that none
of the published work of either could have done, the steps in the
growth of the great generalisation.   We read of the doubts of
one or the other; the gradual accumulation of material facts;
the criticisms and amendments in face of new evidence; and the
slow progress from tentative hypothesis to assured belief.   We
ourselves have grown up since the clash of opinion for and
against the mutability of species died down.   It is hard for us
to understand the strength of the feelings aroused: the bitterness
of the attack by the opponents of the theory, and the fortitude

demanded from its adherents. It is best to obtain evidence on such matters at first hand; and this is what is supplied by the correspondence between Darwin and Hooker.

How complete the understanding between the friends soon became is shown by the provisions made by Darwin for the publication of his manuscripts in case of sudden death. He wrote in August 1854 the definite direction " Hooker by far the best man to edit my species volume": and this notwithstanding that he writes to him as a "stern and awful judge and sceptic." But again, in a letter a few months later, he says to him : " I forgot at the moment that you are the one living soul from whom I have constantly received sympathy." I have already said that Hooker was not only Darwin's first confidant but also the first to accept his theory of mutability of species. But even he did not fully assent to it till after its first publication. The latter point comes out clearly from the letters. In January 1859, six months after the reading of their joint communications to the Linnean Society, Darwin writes to Wallace : " You ask about Lyell's frame of mind. I think he is somewhat staggered, but does not give in...I think he will end by being perverted. Dr Hooker has become almost as heterodox as you or I, and I look at Hooker as by far the most capable judge in Europe." In September 1859 Darwin writes to W. D. Fox : " Lyell has read about half of the volume in clean sheets...He is wavering so much about the immutability of species that I expect he will come round. Hooker has come round, and will publish his belief soon." In the following month, writing to Hooker, Darwin says : " I have spoken of you here as a convert made by me : but I know well how much larger the share has been of your own self-thought." A letter to Wallace of November 1859 bears this postscript : " I think that I told you before that Hooker is a complete convert. If I can convert Huxley I shall be content." And lastly, in a letter to W. B. Carpenter, of the same month, Darwin says: "As yet I know only one believer, but I look at him as of the greatest authority, viz. Hooker." These quotations clearly show that, while Lyell wavered, and Huxley had not yet come in, Hooker was a complete adherent in 1859 to the doctrine of the mutability of species. Excepting

Wallace, he was the first, in fact, of the great group that stood round Darwin, as he was the last of them to survive.

The story of the joint communication of Darwin and of Wallace to the Linnean Society " On the tendency of Species to form Varieties, and on the Perpetuation of Varieties and Species by Natural Means of Selection " will be fresh in the minds of readers, for the fiftieth anniversary of the event was lately cele-brated in London. It was Sir Charles Lyell and Sir Joseph Hooker who jointly communicated the two papers to the society, together with the evidence of the priority of Darwin in the enquiry. Nothing could then have been more apposite than the personal history which Sir Joseph gave at the Darwin-Wallace celebration, held by the Linnean Society in 1908. He then told, at first hand, the exact circumstances under which the joint papers were produced. Nor could the expressions used by the President (Dr Scott) when thanking Sir Joseph, and presenting to him the Darwin-Wallace Medal, have been improved. He said : " The incalculable benefit that your constant friendship, advice, and alliance were to Mr Darwin himself, is summed up in his own words, used in 1864: ' You have represented for many years the whole great public to me.' " The President then added : " Of all men living it is to you more than to any other that the great generalisation of Darwin and Wallace owes its triumph."

The very last appearance of Hooker at any large public gathering of biologists was at the centenary of Darwin's birth, celebrated at Cambridge, in 1909. None who were there will forget the tall figure of the veteran, aged, but still vigorous, with vivacity in every feature. How gladly he accepted the con-gratulations of his many friends, and how heartily he rejoiced over the full acceptance of the theory he had himself done so much to promote. The end came only two years later, in December last. Many will have wished that the great group of the protagonists of Evolution, Darwin, Lyell, and Hooker, should have found their final resting-place together in West-minster Abbey. But this was not to be. Personal and family ties held him closer to Kew. And he lies there in classic ground beside his father.

Having thus sketched the intimate relations which subsisted

between Hooker and Darwin, it remains to appraise his own positive contributions to *Philosophical Biology*. He himself, in his Address as President of the British Association at Norwich in 1868, gives an insight into his early attitude in the enquiry into biological questions. "Having myself," he says, "been a student of Moral Philosophy in a Northern University, I entered on my scientific career full of hopes that Metaphysics would prove a useful mentor, if not a guide in science. I soon found, however, that it availed me nothing, and I long ago arrived at the conclusion so well put by Agassiz, when he says, 'We trust that the time is not distant when it will be universally understood that the battle of the evidences will have to be fought on the field of Physical Science, and not on that of the Metaphysical.'" This was the difficult lesson of the period when Evolution was born. Hooker learned the lesson early. He cleared his mental outlook from all preconceptions, and worked down to the bed-rock of objective fact. Thus he was free to use his vast and detailed knowledge in advancing, along the lines of induction alone, towards sound generalisations. These had their very close relation to questions of the mutability of species. The subject was approached by him through the study of geographical distribution, in which, as we have seen, he had at an early age become the leading authority.

The fame of Sir Joseph Hooker as a Philosophical Biologist rests upon a masterly series of Essays and Addresses. The chief of these were The Introductory Essay to the *Flora Tasmaniae*, dealing with the Antarctic Flora as a whole; The Essay on the Distribution of Arctic Plants, published in 1862; The Discourse on Insular Floras in 1866; The Presidential Address to the British Association at Norwich in 1868; his Address at York, in 1881, on Geographical Distribution; and finally, The Essay on the Vegetation of India, published in 1904. None of these were mere inspirations of the moment. They were the outcome of arduous journeys to observe and to collect, and subsequently of careful analysis of the specimens and of the facts. The dates of publication bear this out. The Essay on the Antarctic Flora appeared about twenty years after the completion of the voyage. The Essay on the Vegetation of India was not published till more than half a century after Hooker first set foot in India. It is upon such foundations that Hooker's reputation as a great constructive thinker is securely based.

The first-named of these essays will probably be estimated as the most notable of them all in the History of Science. It was completed in November 1859, barely a year after the joint communications of Darwin and Wallace to the Linnean Society, and before the *Origin of Species* had appeared. It was to this Essay that Darwin referred when he wrote that "Hooker has come round, and will publish his belief soon." But this publication of his belief was not merely an echo of assent to Darwin's own opinions. It was a reasoned statement, advanced upon the basis of his "own self-thought," and his own wide systematic and geographical experience. From these sources he drew for himself support for the "hypothesis that species are derivative, and mutable." He points out how the natural history of Australia seemed specially suited to test such a theory, on account of the comparative uniformity of the physical features being accompanied by a great variety in its Flora, and the peculiarity of both its Fauna and Flora, as compared with other countries. After the test had been made, on the basis of study of some 8000 species, their characters, their spread, and their relations to those of other lands, he concludes decisively in favour of mutability and a doctrine of progression.

How highly this Essay was esteemed by his contemporaries is shown by the expressions of Lyell and of Darwin. The former writes: "I have just finished the reading of your splendid Essay on the Origin of Species, as illustrated by your wide botanical experience, and think it goes far to raise the variety-making hypothesis to the rank of a theory, as accounting for the manner in which new species enter the world." Darwin wrote: "I have finished your Essay. To my judgment it is by far the grandest and most interesting essay on subjects of the nature discussed I have ever read."

But besides its historical interest in relation to the Species Question, the Essay contained what was up to its time the most scientific treatment of a large area from the point of view of the

Plant-Geographer. He found that the Antarctic, like the Arctic Flora, is very uniform round the Globe. The same species in many cases occur on every island, though thousands of miles of ocean may intervene. Many of these species reappear on the mountains of Southern Chili, Australia, Tasmania, and New Zealand. The Southern Temperate Floras, on the other hand, of South America, South Africa, Australia, and New Zealand differ more among themselves than do the Floras of Europe, Northern Asia, and North America. To explain these facts he suggested the probable former existence, during a warmer period than the present, of a centre of creation of new species in the Southern Ocean, in the form of either a continent or an archipelago, from which the Antarctic Flora radiated. This hypothesis has since been held open to doubt. But the fact that it was suggested shows the broad view which he was prepared to take of the problem before him. His method was essentially that which is now styled " Ecological." Many hold this to be a new phase of botanical enquiry, introduced by Professor Warming in 1895. No one will deny the value of the increased precision which he then brought into such studies. But in point of fact it was Ecology on the grand scale that Sir Joseph Hooker practised in the Antarctic in 1840. Moreover it was pursued, not in regions of old civilisation, but in lands where Nature held her sway untouched by the hand of man.

This Essay on the Flora of the Antarctic was the prototype of the great series. Sir Joseph examined the Arctic Flora from similar points of view. He explained the circumpolar uniformity which it shows, and the prevalence of Scandinavian types, together with the peculiarly limited nature of the Flora of the southward peninsula of Greenland. He extended his enquiries to oceanic islands. He pointed out that the conditions which dictated circumpolar distribution are absent from them; but that other conditions exist in them which account for the strange features which their vegetation shows. He extended the application of such methods to the Himalaya and to Central Asia. He joined with Asa Gray in like enquiries in North America. The latter had already given a scientific explanation of the surprising fact that the plants of the Eastern States resemble more nearly those of China than do those of the Pacific Slope. In resolving these and other problems it was not only the vegetation itself that was studied. The changes of climate in geological time, and of the earth's crust as demonstrated by geologists, formed part of the basis on which he worked. For it is facts such as these which have determined the migration of Floras. And migration, as well as mutability of species, entered into most of his speculations. The Essays of this magnificent series are like pictures painted with a full brush. The boldness and mastery which they show sprang from long discipline and wide experience.

Finally, the chief results of the Phyto-Geographical work of himself and of others were summed up in the great Address on "Geographical Distribution" at York. The Jubilee of the British Association was held there in 1881. It had been decided that each section should be presided over by a past President of the Association, and he had occupied that position at Norwich in 1868. Accordingly at York Hooker was appointed President of the Geographical Section, and he chose as the subject of his Address "The Geographical Distribution of Organic Beings." To him it illustrated "the interdependence of those Sciences which the Geographer should study." It is not enough merely to observe the topography of organisms, but their hypsometrical distribution must also be noted. Further, the changes of area and of altitude in exposed land-surfaces of which geology gives evidence, are essential features in the problem, together with the changes of climate, such as have determined the advance and retrocession of glacial conditions. Having noted these factors, he continued thus: "With the establishment of the doctrine of orderly evolution of species under known laws I close this list of those recognised principles of the science of geographical distri-bution, which must guide all who enter upon its pursuit. As Humboldt was its founder, and Forbes its reformer, so we must regard Darwin as its latest and greatest law-giver." Now, after thirty years, may we not add to these words of his, that Hooker was himself its greatest exponent?

And so we have followed, however inadequately, this great man into the various lines of scientific activity which he pursued.

We have seen him to excel in them all.   The cumulative result is that he is universally held to have been, during several decades, the most distinguished botanist of his time.   He was before all things a philosopher.   In him we see the foremost student of the broader aspects of Plant-Life at the time when evolutionary belief was nascent.   His influence at that stirring period, though quiet, was far-reaching and deep.   His work was both critical and constructive.   His wide knowledge, his keen insight, his fearless judgment were invaluable in advancing that intellectual revolution which found its pivot in the mutability of species. The share he took in promoting it was second only to that of his life-long friend Charles Darwin.

INDEX

Berkeley, as plant pathologist, 231–232
—— as systematic mycologist, 227–230
—— as zoologist, 226
Binney, on Carboniferous plants, 253
—— on coal balls, 245
—— fossil plant tissues and, 307
—— Geological Survey and, 245–246
—— publications of, 246
Blair, on Morison and Ray, 31
Bobart, Jacob, Keeper of Oxford Physic Garden, 17
—— —— the younger, 18
—— —— —— and Morison's work, 18, 23
—— —— —— influence on Ray, 43
Botanic gardens, of British colonies, 136
—— —— of Calcutta, 181–182, 305
—— —— of Cambridge, 153
—— —— Chelsea Physic, 84, 92, 179
—— —— of Glasgow, 129, 130, 292
—— —— Glasnevin, 213
—— —— Government subsidy of, 288
—— —— Kew, 136–137
—— —— Oxford, 17, 18
—— —— purpose of, 281
—— —— rivalry between Edinburgh, 282, 283
Botanical Gazette, Henfrey and, 201, 202
—— *Magazine*, editorship of W. Hooker, 142
—— Society of Edinburgh, founding of, 294
Botanical illustration, Bauer and, 179
—— —— Berkeley and, 226
—— —— Dickson and, 301
—— —— W. Fitch and, 141–142, 246
—— —— Grew and, 52
—— —— Harvey and, 202, 206, 212
—— —— Hill and, 100, 102, 103
—— —— W. Hooker and, 141
—— —— Lindley and, 170, 174
—— —— McGillivray, 244
—— —— Tuffen West and, 200
—— —— Williamson and, 257
Botany, local study of, 249
Botany teaching, Henslow's methods, 158, 159
—— pioneers of, 281, 296
Bottomley, on Gilbert, 233–242
Bower, F. O., 258
—— on W. Hooker, 126–150
—— on J. D. Hooker, 302–323
Boyle, influence on Hales, 66
Bridging species, Ward on, 277
British Algae, Berkeley on, 226
—— Flora, Bentham and Hooker on, 315
—— —— Berkeley on, 227–228
—— —— W. Hooker on, 143
—— —— J. D. Hooker on, 315
Brome grass, Ward on, 276–277

Brongniart, Williamson and, 254–256
—— antiquity of Dicotyledons and, 244
—— on fossil seeds, 257
—— influence of, 246
Broome, collaboration with Berkeley, 229
Brown, Robert, life of, **108–125**
—— —— J. Banks and, 110, 111
—— —— collections of, 112
—— —— Cycad ovule and, 187
—— —— diary of, 111
—— —— Griffith and, 186
—— —— Linnean Society and, 112, 123–124
—— —— on the ovule, 184–185
—— —— period of, 134
—— —— on vegetation of New Holland, 113
Brownian movement, 120
Bryophyta, Griffith and, 188–189, 190
Bud protection, Grew on, 50

Calamites, Williamson on, 253, 254, 256
—— secondary growth in, 254
Calcutta Gardens, Falconer and, 305
—— —— Griffith and, 181–182
Cambridge, Ward at, 262
—— Botanic Gardens, 153
—— Herbarium, Lindley's presentation to, 170
—— Philosophical Society, founding of, 151
—— —— —— presentations to, 152–153
—— —— —— Ward and, 276
—— Professorship, 152, 274
Carboniferous period, Binney on, 246, 253, 307
—— —— J. D. Hooker on, 307
—— —— Williamson on, 4, 253, 256
Cell structure, discovery of, 53
—— —— Grew on, 53–54
—— theory, Williamson and 251
Cesalpino, Andrea, abstract of results of, 12, 13
—— —— classification and, 11
—— —— on Cryptogams, 21
—— —— Morison and, 26
—— —— Theophrastus and, 11
Ceylon Flora, Hooker and, 312, 313
Chelsea Physic Garden, Botany lectures at, 84
—— —— —— Griffith and, 179
—— —— —— Hill's use of, 92
Chemiotaxis, R. Brown on, 115
—— Myoshi on, 275
—— Pfeffer, and, 275
Chlorophyll, Grew's observations on, 59
Circumpolar uniformity, Hooker and, 321
Classification, Adanson on, 41
—— Bauhin and, 14

INDEX

Jackson, Daydon, *Index Kewensis* and, 314
Jodrell Laboratory, Ward at, 262
—— Williamson and, 259
Judd, on Williamson, 251
Jung, Linnaeus and, 15
—— Ray and, 35
—— Systematic botany and, 15
de Jussieu, de Candolle and, 41, 42
—— classification and, 109, 134, 287
—— Tournefort and Ray and, 42

Keeble, F., on Lindley, 164–177
Kew Gardens, Administration of, 309–310
—— —— Bentham's gifts to, 140
—— —— Hales and, 81
—— —— Herbarium at, 305
—— —— Herbarium and Library at, 141, 310
—— —— Hill and, 93
—— —— W. Hooker Director of, 127, 130, 133–4, 136–7, 140, 149
—— —— the two Hookers at, 308–309
—— —— Lindley and, 136–138, 169–170
—— —— Mycological Herbarium at, 231
—— —— Orchid Herbarium at, 170
—— —— Thiselton-Dyer and, 81, 150, 203, 259
Kidston, Witham's collection and, 245
Kingia, R. Brown on, 117
King's Botanist, Alston as, 283
—— —— Arthur as, 284
Knaut, work of, 26, 39

Lang, on Griffith, 178–191
Lankester, E., 193
Lankester, Ray, on *Schizomycetes*, 265
Lawes, agriculture and, 235
—— Gilbert and, 235–236
—— on nitrogen assimilation, 267
Leaves, classification on form of, 21
—— movements of, 96–99
—— structure and functions of, 96–97
—— vernation of, 50
Leguminous nodules, Gilbert on, 239
—— —— Hellriegel and, 240
—— —— Ward on, 267
Lepidodendron, Williamson and, 258
Lichens, Ward on, 264, 270
Liebig, mineral theory of, 237–240
—— controversy with Gilbert, 238
Light on plants, Hales and, 80, 81
—— —— Hill on, 97
Lily disease, Ward on, 266, 267
Lindley, John, life of, **164–177**
—— —— activities of, 166
—— —— J. Banks and, 168
—— —— *Botanical Register* and, 174

Lindley, John, characteristics of, 177
—— —— on cryptogams, 195
—— —— on Darwin, 174
—— —— Horticulture and, 171
—— —— Hutton and, 245, 248
—— —— Library and, 174, 177
—— —— literary work of, 169–174
—— —— Professorship of, 168, 169
—— —— Williamson and, 250
Linnaeus, British botany and, 6
—— Hope and, 286
—— Jung and, 15
—— on Morison and Cesalpino, 27–28, 42, 43
—— on Ray, 38, 42, 43
—— Taxonomy and, 6
—— method, 2, 3, 39
—— Adanson and, 41
—— Hill and, 101–103
—— England and, 39, 40
—— period, 109, 134
—— system, Alston on, 285
—— Hope on, 289
—— influence of, 193
Linnean school, influence of, 195
—— Society, R. Brown and, 111, 112, 114, 121–124
—— —— Berkeley and, 232
—— —— Darwin-Wallace Celebration, 317, 318
—— —— Griffith and, 183–185
—— —— Harvey and, 207
—— —— Ward and, 263, 264
—— —— publications of, 184–186, 253
London University, Botany teaching at, 179
—— —— Henslow and, 154
Loranthaceae, Griffith on, 185–186
Lotsy, morphology and, 186
Lycopods, Williamson and, 256
Lyell, *Calamites* and, 253
—— J. H. Hooker and, 308, 320
—— mutability of species and, 317
Lyginodendron, Williamson and, 256, 257

Malpighi, anatomy and, 44, 135, 286
—— Grew and, 6, 48, 63
—— Hales and, 67, 68, 81
—— on seeds and seedlings, 35, 36
Manchester, Geological Society of, 245
—— Natural History Society of, 249
—— Professorship at, 250
—— Ward and, 264
Mangroves, Griffith and, 186–187
Manures, experiments with, 234, 236, 237
—— effect of, 237
—— nitrogenous, 238
Marchantia, Henfrey on, 199
Massee, George, on Berkeley, 225–232
Materia Medica, Alston and, 285

Printed in the United States
By Bookmasters

Printed in the United States
By Bookmasters